# Introduction to Mathematical Thinking

# Keith Devlin

Publisher:
Keith Devlin
331 Poe St, Unit 4
Palo Alto, CA 94301
USA
http://profkeithdevlin.com

**Publication data**

Devlin, Keith, Introduction to Mathematical Thinking

First published, July 2012

ISBN-13: 978-0615653631

ISBN-10: 0615653634

# Contents

# Preface

Many students encounter difficulty going from high school math to college-level mathematics. Even if they do well at math in school, most students are knocked off course for a while by the shift in emphasis from the K-12 focus on mastering procedures to the "mathematical thinking" characteristic of much university mathematics. Though the majority survive the transition, many do not, and leave mathematics for some other major (possibly outside the sciences or other mathematically-dependent subjects). To help incoming students make the shift, colleges and universities often have a "transition course."

This short book is written to accompany such a course, but it is not a traditional "transition textbook." Rather than give beginning college students (and advanced high school seniors) a crash course in mathematical logic, formal proofs, some set theory, and a bit of elementary number theory and elementary real analysis, as is commonly done, I attempt to help students develop that crucial but elusive ability: *mathematical thinking*. This is not the same as "doing math," which usually involves the application of procedures and some heavy-duty symbolic manipulations. Mathematical thinking, by contrast, is a specific way of thinking about things in the world. It does not have to be about mathematics at all, though I would argue that certain parts of mathematics provide the ideal contexts for learning how to think that way, and in this book I will concentrate my attention on those areas.

Mathematicians, scientists, and engineers need to "do math." But for life in the twenty-first century, everyone benefits from being able to think mathematically to some extent. (Mathematical thinking includes logical and analytic thinking as well as quantitative reasoning, all crucial abilities.) This is why I have tried to make this book accessible to anyone who wants or needs to extend and improve their analytic thinking skills. For the student who goes beyond a basic grasp of logical and analytic thinking, and truly masters mathematical thinking, there is a payoff at least equal to those advantages incidental to twenty-first century citizenship: mathematics goes from being confusing, frustrating, and at times seemingly impossible, to making sense and being hard but *doable*.

I developed one of the first college transition courses in the late 1970s, when I was teaching at the University of Lancaster in England, and I wrote one of the first transition textbooks, *Sets, Functions and Logic*, which was published in 1981.[1] Today, when I give such a course, I structure it differently, giving it the broader focus of "mathematical thinking," and likewise this book is different from my earlier

---

[1] Now in its third edition: *Sets, Functions, and Logic: An Introduction to Abstract Mathematics*, Chapman & Hall, CRC Mathematics Series.

one.[2] While I understand the rationale behind the more familiar transition courses and textbooks, the course I give today, and along with it this book, seek to serve a much wider audience. (In particular, I no longer devote time to formal, mathematical logic. While logic provides a useful *model* of mathematical reasoning—which is the reason it was developed in the first place—I no longer think it is the best way to develop practical, logical reasoning skills.) In adopting this broader, societal perspective, I believe my course—and this book—will not only still help beginning college and university mathematics students successfully negotiate the transition from high school, it will also help anyone develop better reasoning skills.

For some reason, transition course textbooks are generally hugely expensive, in some cases over $100, which is a lot for a book that likely will be used for at most one semester. This book is designed to accompany transition courses that last only five to seven weeks. For that reason, I decided to self-publish it as a low-cost, print-on-demand book. I did however engage an experienced, professional mathematics textbook editor, Joshua D. Fisher, to go over my entire manuscript prior to publication. The final form of the book owes a lot to his expertise, and for that I am very appreciative.

Keith Devlin
Stanford University
July 2012

---

[2]Because my earlier book and this new one both arise from transition courses I have developed, there is still considerable overlap between the two books, as indeed there is between my two books and transition books by other authors, but this book has a different focus and is targeted at a different, and broader, audience than all the rest.

# What this book is about

Dear reader,

I wrote this book with two kinds of reader in mind: (1) the high school graduate entering college or university who wants to (or could) major in mathematics or some math-dependent subject and (2) anyone who, for whatever reason, wants or needs to develop or improve their analytic thinking skills. Regardless, the focus is on learning to think a certain (very powerful) way.

You won't learn any mathematical procedures from this book, nor will you have to apply any! Though the final chapter focuses on numbers (elementary number theory and the basics of real analysis), the amount of "traditional" mathematical material I cover on those topics is tiny. The last chapter simply provides excellent examples that have helped mathematicians themselves develop, over time, the analytic thinking skills I shall describe throughout this book.

During the nineteenth century, the need for those analytic thinking skills among the wider, general population grew with increasing democratization and "flattening" of society, which gave—and continue to give—every citizen more and more freedom and opportunities to play a significant, self-directed role in business or society. Today, more than ever, good analytic thinking skills are essential for anyone who wants to take full advantage of the opportunities for self-growth and advancement that contemporary democratic societies offer.

I have been teaching and writing books about the patterns of thinking required to succeed in college-level[3] (pure) mathematics for several decades. Yet, it was only in the last fifteen years, as I found myself doing a fair amount of consulting work with industry and government, that I heard first hand that the "mathematical thinking skills" that were the focus of those courses and books were exactly what business and government leaders said they valued most in many of their employees. Rarely would a CEO or the head of a government lab say that they wanted people with specific skills; rather, their need was for people who had good analytic thinking skills and were able to acquire new specific skills when needed.

As a consequence of the ideas borne out of these divergent, though obviously interconnected experiences—in academia and business—in writing this new book I have, for the first time, tried to structure the development in a way that makes it accessible to a wide audience.

With that said, the remainder of this introduction is directed primarily at the entering (or about-to-enter) college student faced with having to take some (pure) mathematics courses. For the more general reader, the value of what I will say is

---

[3]I shall use the word "college" to mean "college or university" throughout.

that the mathematical thinking skills required to master modern pure mathematics are precisely the crucial mental abilities required to succeed in many professions and walks of life, as I just discussed.

* * *

Dear student,

As you are about to discover, the transition from high school mathematics to college-level (pure) abstract mathematics is a difficult one. Not because the mathematics gets harder. Those students who have successfully made the transition will likely agree that college math feels in many ways easier. What causes the problem for many—as I mentioned above—is only the change in emphasis. In high school, the focus is primarily on mastering procedures to solve various kinds of problems. This gives the process of learning very much a flavor of reading and absorbing the recipes of a kind of mathematical cookbook. At college, the focus is on *learning to think in a different, specific way*—to think like a mathematician.

(Actually, this is not true of all college math courses. Those courses designed for science and engineering students are often very much in the same vein as the calculus courses that typically form the pinnacle of high school mathematics. It's the courses that compose the bulk of the mathematics major that are different. Since some of those courses are usually required for more advanced work in science and engineering, students in those disciplines also may find themselves faced with this "different kind" of math.)

Thinking mathematically isn't a different kind of math, but rather a broader and more up-to-date—though not more dilute—perspective on mathematics. The high school curriculum typically focuses on mathematical procedures and largely ignores the rest of the subject. Still, to the student—you—college math does initially feel like a completely different discipline. It certainly did when I started my undergraduate studies in math. If you go to college to study math (or a math-heavy subject like physics), you must have done pretty well in math at school. That means, you got to be good at mastering and following procedures (and to some extent doing so under time constraints). And that is what the school system rewarded you for. Then you head off to college and all the rules change. In fact, at first it feels as if there are no rules, or if they are the professors are keeping them secret.

Why the change in emphasis when you get to college? Simple. Education is about learning new skills and increasing your capacity for doing things. Once you have shown you can learn new mathematical procedures, which you have by the time you graduate from high school, there is little to be gained from being taught more of the same. You will be able to pick up new techniques whenever you need them.

For instance, once a piano student has mastered one Tchaikovsky concerto, with at most a bit of practice—but essentially no new teaching—they'll be able to play another. From then on, the student's focus should be on expanding their repertoire to include other composers, or to understand music sufficiently well to compose their own.

Analogously, in the case of math, your goal at college is to develop the thinking skills that will allow you to solve novel problems (either practical, real-world problems or ones that arise in math or science) for which you don't know a standard procedure. In some cases, there may not be a standard procedure. (This was the case when the Stanford graduate students Larry Page and Serjey Brin developed a new mathematical procedure to search for information, leading to their creation of *Google*.)

To put it another way (one that makes it clear why mathematical thinking is so valuable in the modern world), before college you succeed in math by learning to "think inside the box"; at college, success in math comes from learning to "think outside the box," an ability that practically every major employer today says they value highly in their workers.

The primary focus of this book, like all "transition books" and "transition courses," is on helping you learn how to approach a new problem, one that does not quite fit any template you are familiar with. It comes down to learning how to *think* (about a given problem).

The first key step (there are two) you have to make in order to successfully negotiate this school-to-college transition, is to learn to stop looking for a formula to apply or a procedure to follow. Approaching a new problem by looking for a template—say a worked example in a textbook or presented on a *YouTube* video— and then just changing the numbers, usually won't work. (Working that way is still useful in many parts of college math, and for real-world applications, so all that work you did at high school won't go to waste. But it isn't enough for the new kind of "mathematical thinking" you'll be required to do in many of your college math courses.)

So, if you can't solve a problem by looking for a template to follow, a formula to plug some numbers into, or a procedure to apply, what do you do? The answer— and this is the second key step—is you think about the problem. Not the form it has (which is probably what you were taught to do at school, and which served you well there), but what it actually says. That sounds as though it ought to be easy, but most of us initially find it extremely hard and very frustrating. Given that this will likely be your experience as well, it might help you to know that there is a good reason for the change. It has to do with (real-world) applications of mathematics: I'll elaborate in Chapter 1, but for now, I'll just give you an analogy.

If we compare mathematics with the automotive world, school math corresponds to learning to drive. In the automotive equivalent to college math, in contrast, you learn how a car works, how to maintain and repair it, and, if you pursue the subject far enough, how to design and build your own car.

I'll finish this brief introduction with some pointers to keep in mind as you work through this book.

- The only prerequisite for this book is completion (or pending completion) of a typical high school mathematics curriculum. In one or two places (particularly in the final chapter) I assume some knowledge of elementary set theory (primarily the notions and properties of set-inclusion, unions, and intersections). I include the requisite material as an appendix for anyone not already familiar with the subject.

- Bear in mind that one reason you are likely to find the going tough is that everything will seem unmotivated. The goal is to provide you with the foundation on which to build the mathematical thinking that comes later—mathematics that you do not yet know about! There is no avoiding an element of bootstrapping the new way of thinking.

- Make your focus *understanding* the new concepts and ideas.

- Don't rush. There are very few new facts to learn—note how thin this book is—but a lot to comprehend!

- Try the exercises—as many of them as possible. They are included to aid your understanding.

- Discuss any difficulties which arise with your colleagues and with your instructor. Few of us master a crucial shift in thinking on our own.

- I should stress that this is not a textbook designed for self-study. It is written as a course companion, something to supplement human instruction and to be consulted whenever you feel you need supplementary information from a source other than your instructor.

- There are many exercises throughout the book. I strongly urge you to do them. They are an integral part of the book. Unlike textbooks, however, I have not provided answers to the exercises. That is not an oversight, but a deliberate choice on my part. Learning to think mathematically is not about getting answers. (Though once you have learned how to think mathematically, getting the right answer becomes a lot easier than when you were just following procedural recipes.) If you want to know if you have got something right—and we all do—you should seek out someone who knows. Deciding whether a piece of mathematical reasoning is correct is a value judgment that requires expertise. Students not infrequently get what on the surface looks like the right answer but on closer examination turns out to be wrong. Sure, there are some exercises where I could have safely given the answer, but I also wanted to reinforce the crucial message that negotiating the school-to-university math transition is all about the *process*—about trying and reflecting—not "getting answers."

- If at all possible, work with others. Working alone is common at high school, with its focus on *doing*, but mastering transition material is all about thinking, and discussing your work with others is a much better approach than solitary study. Analyzing and critiquing attempts at proofs by fellow students will greatly assist your own learning and comprehension.

- Do not try to rush through any section, even if at first glance it looks easy.[4] This entire book consists of *basic* material required elsewhere (indeed practically *everywhere*) in college-level mathematics. Everything you will find in this book is included because it generally causes problems for the beginner. (Trust me on this.)

- Don't give up. Students all around the world managed it last year, and the year before that. So did I, many years ago. So will you!

- Oh yes, one more thing: Don't rush.

- Remember, the goal is understanding and developing a new way of thinking— one that you will find valuable in many walks of life.

- School math was about *doing*; college math is largely about *thinking*.

- Three final words of advice. Take. Your. Time.

Good luck. :-)

Keith Devlin
Stanford University
July 2012

---

[4]Yes, I know I said this a mere six short paragraphs ago. The repetition is deliberate. It's an important point.

# 1

# What is mathematics?

For all the time schools devote to the teaching of mathematics, very little (if any) is spent trying to convey just what the subject is about. Instead, the focus is on learning and applying various procedures to solve math problems. That's a bit like explaining soccer by saying it is executing a series of maneuvers to get the ball into the goal. Both accurately describe various key features, but they miss the what and the why of the big picture.

Given the demands of the curriculum, I can understand how this happens, but I think it is a mistake. Particularly in today's world, a general understanding of the nature, extent, power, and limitations of mathematics is valuable to any citizen.[1] Over the years, I've met many people who graduated with degrees in such mathematically rich subjects as engineering, physics, computer science, and even mathematics itself, who have told me that they went through their entire school and college-level education without ever gaining a good overview of what constitutes modern mathematics. Only later in life do they sometimes catch a glimpse of the true nature of the subject and come to appreciate its pervasive role in modern life.[2]

## 1.1   More than arithmetic

Most of the mathematics used in present-day science and engineering is no more than three- or four-hundred years old, much of it less than a century old. Yet the typical high school curriculum comprises mathematics at least that old—some of it over two-thousand years old!

Now, there is nothing wrong with teaching something so old. As the saying goes, if it ain't broke, don't fix it. The algebra that the Arabic-speaking traders developed in the eighth and ninth centuries (the word *algebra* comes from the Arabic term *al-jabr*, meaning "restoration" or "reunion of broken parts") to increase efficiency in their business transactions remains as useful and important today as it was then, even though today we may now implement it in a spreadsheet macro rather than by medieval finger calculation. But time moves on and society advances. In the

---

[1] If you have not yet done so, please go back and read the Introduction to this book. It is very relevant here, and throughout the book.

[2] See the previous footnote.

process, the need for new mathematics arises and, in due course, is met. Education needs to keep pace.

Mathematics can be said to have begun with the invention of numbers and arithmetic, which is believed to have occurred around ten thousand years ago, with the introduction of money. (Yes, apparently it began with money!)

Over the ensuing centuries, the ancient Egyptians and Babylonians expanded the subject to include geometry and trigonometry.[3] In those civilizations, mathematics was largely utilitarian, and very much of a "cookbook" variety. ("Do such and such to a number or a geometric figure, and you will get the answer.")

The period from around 500 BCE to 300 CE was the era of Greek mathematics. The mathematicians of ancient Greece had a particularly high regard for geometry. Indeed, they treated numbers geometrically, as measurements of length, and when they discovered that there were lengths to which their numbers did not correspond (essentially the discovery of irrational numbers), their study of number largely came to a halt.[4]

In fact, it was the Greeks who made mathematics into an area of study, not merely a collection of techniques for measuring, counting, and accounting. Around 500 BCE, Thales of Miletus (Miletus is now part of Turkey) introduced the idea that the precisely stated assertions of mathematics could be logically proved by formal arguments. This innovation marked the birth of the theorem, now the bedrock of mathematics. This formal approach by the Greeks culminated in the publication of Euclid's *Elements*, reputedly the most widely circulated book of all time after the Bible.[5]

By and large, school mathematics is based on all the developments I listed above, together with just two further advances, both from the seventeenth century: calculus and probability theory. Virtually nothing from the last three hundred years has found its way into the classroom. Yet most of the mathematics used in today's world was developed in the last two hundred years, let alone the last three hundred!

As a result, anyone whose view of mathematics is confined to what is typically taught in schools is unlikely to appreciate that research in mathematics is a thriving, worldwide activity, or to accept that mathematics permeates, often to a considerable extent, most walks of present-day life and society. For example, they are unlikely to know which organization in the United States employs the greatest number of Ph.D.s in mathematics. (The answer is almost certainly the National Security Agency, though the exact number is an official secret. Most of those mathematicians work on code breaking, to enable the agency to read encrypted messages that are intercepted by monitoring systems—at least, that is what is generally

---

[3]Other civilizations also developed mathematics; for example, the Chinese and the Japanese. But the mathematics of those cultures does not appear to have had a direct influence on the development of modern western mathematics, so in this book I will ignore them.

[4]There is an oft repeated story that the young Greek mathematician who made this discovery was taken out to sea and drowned, lest the awful news of what he had stumbled upon should leak out. As far as I know, there is no evidence whatsoever to support this fanciful tale. Pity, since it's a great story.

[5]Given today's mass market paperbacks, the definition of "widely circulated" presumably has to incorporate the number of years the book has been in circulation.

assumed, though again the Agency won't say. Though most Americans probably know that the NSA engages in code breaking, many do not realize it requires mathematics, and hence do not think of the NSA as an organization that employs a large number of advanced mathematicians.)

The explosion of mathematical activity that has taken place over the past hundred years or so in particular has been dramatic. At the start of the twentieth century, mathematics could reasonably be regarded as consisting of about twelve distinct subjects: arithmetic, geometry, calculus, and several more. Today, the number of distinct categories is somewhere between sixty and seventy, depending how you count them. Some subjects, like algebra or topology, have split into various subfields; others, such as complexity theory or dynamical systems theory, are completely new areas of study.

The dramatic growth in mathematics led in the 1980s to the emergence of a new definition of mathematics as the *science of patterns*. According to this description, the mathematician identifies and analyzes abstract patterns—numerical patterns, patterns of shape, patterns of motion, patterns of behavior, voting patterns in a population, patterns of repeating chance events, and so on. Those patterns can be either real or imagined, visual or mental, static or dynamic, qualitative or quantitative, utilitarian or recreational. They can arise from the world around us, from the pursuit of science, or from the inner workings of the human mind. Different kinds of patterns give rise to different branches of mathematics. For example:

- Arithmetic and number theory study the patterns of number and counting.

- Geometry studies the patterns of shape.

- Calculus allows us to handle patterns of motion.

- Logic studies patterns of reasoning.

- Probability theory deals with patterns of chance.

- Topology studies patterns of closeness and position.

- Fractal geometry studies the self-similarity found in the natural world.

## 1.2 Mathematical notation

One aspect of modern mathematics that is obvious to even the casual observer is the use of abstract notations: algebraic expressions, complicated-looking formulas, and geometric diagrams. The mathematicians' reliance on abstract notation is a reflection of the abstract nature of the patterns they study.

Different aspects of reality require different forms of description. For example, the most appropriate way to study the lay of the land or to describe to someone how to find their way around a strange town is to draw a map. Text is far less appropriate. Analogously, annotated line drawings (blueprints) are most appropriate for representing the construction of a building. And musical notation is most appropriate for representing music on paper. In the case of various kinds of abstract, formal patterns and abstract structures, the most appropriate means of description and analysis is mathematics, using mathematical notations, concepts, and procedures.

For example, the commutative law for addition could be written in English as:

*When two numbers are added, their order is not important.*

However, it is usually written in the symbolic form:

$$m + n = n + m$$

While the symbolic form has no significant advantage for a simple example such as this, such is the complexity and the degree of abstraction of the majority of mathematical patterns, that to use anything other than symbolic notation would be prohibitively cumbersome. And so the development of mathematics has involved a steady increase in the use of abstract notations.

Though the introduction of symbolic mathematics in its modern form is generally credited to the French mathematician Françoise Viète in the sixteenth century, the earliest appearance of algebraic notation seems to have been in the work of Diophantus, who lived in Alexandria some time around 250 CE. His thirteen volume treatise *Arithmetica* (only six volumes have survived) is generally regarded as the first algebra textbook. In particular, Diophantus used special symbols to denote the unknown in an equation and to denote powers of the unknown, and he had symbols for subtraction and for equality.

These days, mathematics books tend to be awash with symbols, but mathematical notation no more is mathematics than musical notation is music. A page of sheet music *represents* a piece of music; the music itself is what you get when the notes on the page are sung or performed on a musical instrument. It is in its performance that the music comes alive and becomes part of our experience. The music exists not on the printed page but in our minds. The same is true for mathematics. The symbols on a page are just a *representation* of the mathematics. When read by a competent performer (in this case, someone trained in mathematics), the symbols on the printed page come alive—the mathematics lives and breathes in the mind of the reader like some abstract symphony.

To repeat, the reason for the abstract notation is the abstract nature of the patterns that mathematics helps us identify and study. For example, mathematics is essential to our understanding the invisible patterns of the universe. In 1623, Galileo wrote,

> The great book of nature can be read only by those who know the language in which it was written. And this language is mathematics.[6]

In fact, physics can be accurately described as the universe seen through the lens of mathematics.

---

[6] *The Assayer*. This is an oft repeated paraphrase of his actual words.

To take just one example, as a result of applying mathematics to formulate and understand the laws of physics, we now have air travel. When a jet aircraft flies overhead, you can't see anything holding it up. Only with mathematics can we "see" the invisible forces that keep it aloft. In this case, those forces were identified by Isaac Newton in the seventeenth century, who also developed the mathematics required to study them, though several centuries were to pass before technology had developed to a point where we could actually use Newton's mathematics (enhanced by a lot of additional mathematics developed in the interim) to build airplanes. This is just one of many illustrations of one of my favorite memes for describing what mathematics does: *mathematics makes the invisible visible.*

## 1.3 Modern college-level mathematics

With that brief overview of the historical development of mathematics under our belts, I can start to explain how modern college math came to differ fundamentally from the math taught in school.

Up to about 150 years ago, although mathematicians had long ago expanded the realm of objects they studied beyond numbers (and algebraic symbols for numbers), they still regarded mathematics as primarily about *calculation*. That is, proficiency at mathematics essentially meant being able to carry out calculations or manipulate symbolic expressions to solve problems. By and large, high school mathematics is still very much based on that earlier tradition.

But during the nineteenth century, as mathematicians tackled problems of ever greater complexity, they began to discover that these earlier intuitions about mathematics were sometimes inadequate to guide their work. Counterintuitive (and occasionally paradoxical) results made them realize that some of the methods they had developed to solve important, real-world problems had consequences they could not explain. The Banach–Tarski Paradox, for example, says you can, in principle, take a sphere and cut it up in such a way that you can reassemble it to form two identical spheres each the same size as the original one. Because the mathematics is correct, the Banach–Tarski result had to be accepted as a fact, even though it defies our imagination.

It became clear, then, that mathematics can lead to realms where the only understanding is through the mathematics itself. In order to be confident that we can rely on discoveries made by way of mathematics—but not verifiable by other means—mathematicians turned the methods of mathematics inwards, and used them to examine the subject itself.

This introspection led, in the middle of the nineteenth century, to the adoption of a new and different conception of mathematics, where the primary focus was no longer on performing calculations or computing answers, but formulating and understanding abstract concepts and relationships. This was a shift in emphasis from *doing* to *understanding*. Mathematical objects were no longer thought of as given primarily by formulas, but rather as carriers of conceptual properties. Proving something was no longer a matter of transforming terms in accordance with rules, but a process of logical deduction from concepts.

This revolution—for that is what it amounted to—completely changed the way mathematicians thought of their subject. Yet, for the rest of the world, the shift may as well have not occurred. The first anyone other than professional mathematicians knew that something had changed was when the new emphasis found its way into the undergraduate curriculum. If you, as a college math student, find yourself reeling after your first encounter with this "new math," you can lay the blame at the feet of the mathematicians Lejeune Dirichlet, Richard Dedekind, Bernhard Riemann, and all the others who ushered in the new approach.

As a foretaste of what is to come, I'll give one example of the shift. Prior to the nineteenth century, mathematicians were used to the fact that a formula such as $y = x^2 + 3x - 5$ specifies a *function* that produces a new number $y$ from any given number $x$. Then the revolutionary Dirichlet came along and said to forget the formula and concentrate on what the function *does* in terms of input–output behavior. A *function*, according to Dirichlet, is any rule that produces new numbers from old. The rule does not have to be specified by an algebraic formula. In fact, there's no reason to restrict your attention to numbers. A function can be any rule that takes objects of one kind and produces new objects from them.

This definition legitimizes functions such as the one defined on real numbers by the rule:

If $x$ is rational, set $f(x) = 0$; if $x$ is irrational, set $f(x) = 1$.

Try graphing that monster!

Mathematicians began to study the properties of such *abstract* functions, specified not by some formula but by their behavior. For example, does the function have the property that when you present it with different starting values it always produces different answers? (This property is called *injectivity*.)

This abstract, conceptual approach was particularly fruitful in the development of the new subject called real analysis, where mathematicians studied the properties of continuity and differentiability of functions as abstract concepts in their own right. French and German mathematicians developed the "epsilon-delta definitions" of continuity and differentiability that to this day cost each new generation of post-calculus mathematics students so much effort to master.

Again, in the 1850s, Riemann defined a complex function *by its property of differentiability*, rather than a formula, which he regarded as secondary.

The *residue classes* defined by the famous German mathematician Karl Friedrich Gauss (1777–1855), which you are likely to meet in an algebra course, were a forerunner of the approach—now standard—whereby a mathematical structure is defined as a set endowed with certain operations, whose behaviors are specified by axioms.

Taking his lead from Gauss, Dedekind examined the new concepts of *ring*, *field*, and *ideal*—each of which was defined as a collection of objects endowed with certain operations. (Again, these are concepts you are likely to encounter soon in your post-calculus mathematics education.)

And there were many more changes.

Like most revolutions, the nineteenth century change had its origins in times long before the main protagonists came on the scene. The Greeks had certainly shown an interest in mathematics as a conceptual endeavor, not just calculation; and in the seventeenth century, calculus co-inventor Gottfried Leibniz thought deeply about both approaches. But for the most part, until the nineteenth century, mathematics was viewed primarily as a collection of procedures for solving problems. To today's mathematicians, however, brought up entirely with the post-revolutionary conception of mathematics, what in the nineteenth century was a revolution is simply taken to be what mathematics is. The revolution may have been quiet, and to a large extent forgotten, but it was complete and far reaching. And it sets the scene for this book, the main aim of which is to provide you with the basic mental tools you will need to enter this new world of modern mathematics (or at least to learn to think mathematically).

Although the post-nineteenth-century conception of mathematics now dominates the field at the post-calculus, college level, it has not had much influence on high school mathematics—which is why you need a book like this to help you make the transition. There was one attempt to introduce the new approach into school classrooms, but it went terribly wrong and soon had to be abandoned. This was the so-called "New Math" movement of the 1960s. What went wrong was that by the time the revolutionaries' message had made its way from the mathematics departments of the leading universities into the schools, it was badly garbled.

To mathematicians before and after the mid-1800s, both calculation and understanding had always been important. The nineteenth century revolution merely shifted the *emphasis* regarding which of the two the subject was really about and which played the derivative or supporting role. Unfortunately, the message that reached the nation's school teachers in the 1960s was often, "Forget calculation skill, just concentrate on concepts." This ludicrous and ultimately disastrous strategy led the satirist (and mathematician) Tom Lehrer to quip, in his song *New Math*, "It's the method that's important, never mind if you don't get the right answer." After a few sorry years, "New Math" (which was already over a hundred years old, note) was largely dropped from the school syllabus.

Such is the nature of educational policy-making in free societies that it is unlikely such a change could ever be made in the foreseeable future, even if it were done properly the second time around. It's also not clear (at least to me) that such a change would be altogether desirable. There are educational arguments (which in the absence of hard evidence either way are hotly debated) that say the human mind has to achieve a certain level of mastery of computation with abstract mathematical entities before it is able to reason about their properties.

## 1.4  Why do you have to learn this stuff?

It should be clear by now that the nineteenth century shift from a computational view of mathematics to a conceptual one was a change within the professional mathematical community. Their interest, as professionals, was in the very nature of mathematics. For most scientists, engineers, and others who make use of mathematical methods in their daily work, things continued much as before, and that remains the same today. Computation (and getting the right answer) remains just as important as ever, and even more widely used than at any time in history.

As a result, to anyone outside the mathematical community, the shift looks more like an *expansion* of mathematical activity than a change of focus. Instead of just learning procedures to solve problems, college-level math students today *also* (i.e., *in addition*) are expected to master the underlying concepts and be able to justify the methods they use.

Is it reasonable to require this? Although professional mathematicians—whose job it is to develop new mathematics and certify its correctness—need such conceptual understanding, why make it a requirement for those whose goal is to pursue a career in which mathematics is merely a tool? (Engineering for example.)

There are two answers, both of which have a high degree of validity. (SPOILER: It only appears that there are two answers. On deeper analysis, they turn out to be the same.)

First, education is not solely about the acquisition of specific tools to use in a subsequent career. As one of the greatest creations of human civilization, mathematics should be taught alongside science, literature, history, and art in order to pass along the jewels of our culture from one generation to the next. We humans are far more than the jobs we do and the careers we pursue. Education is a preparation for life, and only part of that is the mastery of specific work skills.

That first answer should surely require no further justification. The second answer addresses the tools-for-work issue.

There is no question that many jobs require mathematical skills. Indeed, in most industries, at almost any level, the mathematical requirements turn out to be higher than is popularly supposed, as many people discover when they look for a job and find their math background lacking.

Over many years, we have grown accustomed to the fact that advancement in an industrial society requires a workforce that has mathematical skills. But if you look more closely, those skills fall into two categories. The first category comprises people who, given a mathematical problem (i.e., a problem already formulated in mathematical terms), can find its mathematical solution. The second category comprises people who can take a new problem, say in manufacturing, identify and describe key features of the problem mathematically, and use that mathematical description to analyze the problem in a precise fashion.

In the past, there was a huge demand for employees with Type 1 skills, and a small need for Type 2 talent. Our mathematics education process largely met both needs. It has always focused primarily on producing people of the first variety, but some of them inevitably turned out to be good at the second kind of activities as well. So all was well. But in today's world, where companies must constantly innovate to stay in business, the demand is shifting toward Type 2 mathematical thinkers—to people who can think outside the mathematical box, not inside it. Now, suddenly, all is not well.

There will always be a need for people with mastery of a range of mathematical techniques, who are able to work alone for long periods, deeply focused on a specific mathematical problem, and our education system should support their development. But in the twenty-first century, the greater demand will be for Type 2 ability. Since we don't have a name for such individuals ("mathematically able" or even "mathematician" popularly imply Type 1 mastery), I propose to give them one: *innovative mathematical thinkers.*

This new breed of individuals (well, it's not new, I just don't think anyone has shone a spotlight on them before) will need to have, above all else, a good conceptual understanding of mathematics, its power, its scope, when and how it can be applied, and its limitations. They will also have to have a solid mastery of some basic mathematical skills. But their skills mastery does not have to be stellar. A far more important requirement is that they can work well in teams, often cross-disciplinary teams, they can see things in new ways, they can quickly learn and come up to speed on a new technique that seems to be required, and they are very good at adapting old methods to new situations.

How do we educate such individuals? We concentrate on the conceptual thinking that lies behind all the specific techniques of mathematics. Remember that old adage, "Give a man a fish, he'll eat for a day. Teach a man to fish, he'll eat for a lifetime."? It's the same for mathematics education for twenty-first century life. There are so many different mathematical techniques, with new ones being developed all the time, that it is impossible to cover them all in K-16 education. By the time a college frosh graduates and enters the workforce, many of the specific techniques learned in those college-years are likely to be no longer as important, while new ones are all the rage. The educational focus has to be on learning how to learn.

The increasing complexity in mathematics led mathematicians in the nineteenth century to shift (broaden, if you prefer) the focus from computational skills to the underlying, foundational, conceptual thinking ability. Now, 150 years later, the changes in society that were facilitated in part by that more complex mathematics, have made that focal shift important not just for professional mathematicians but for everyone who learns math with a view to using it in the world.

So now you know not only why mathematicians in the nineteenth century shifted the focus of mathematical research, but also why, from the 1950s onwards, college mathematics students were expected to master conceptual mathematical thinking as well. In other words, you now know why your college or university wants you to take that transition course, and perhaps work your way through this book. Hopefully, you also now realize why it can be important to YOU in living your life, beyond the immediate need of surviving your college math courses.

# 2

# Getting precise about language

The American Melanoma Foundation, in its 2009 Fact Sheet, states that:

*One American dies of melanoma almost every hour.*

To a mathematician, such a claim inevitably raises a chuckle, and occasionally a sigh. Not because mathematicians lack sympathy for a tragic loss of life. Rather, if you take the sentence literally, it does not at all mean what the AMF intended. What the sentence actually claims is that there is one American, Person X, who has the misfortune—to say nothing of the remarkable ability of almost instant resurrection—to die of melanoma every hour. The sentence the AMF writer should have written is

*Almost every hour, an American dies of melanoma.*

Such misuse of language is fairly common, so much so that arguably it is not really misuse. Everyone reads the first sentence as having the meaning captured accurately by the second. Such sentences have become figures of speech. Apart from mathematicians and others whose profession requires precision of statements, hardly anyone ever notices that the first sentence, when read literally, actually makes an absurd claim.

When writers and speakers use language in everyday contexts to talk about everyday circumstances, they and their readers and listeners almost always share a common knowledge of the world (and in particular what is being written or spoken about), and that common knowledge can be drawn upon to determine the intended meaning. But when mathematicians (and scientists) use language in their work, there often is limited or no shared, common understanding—everyone is engaged in a process of discovery. Moreover, in mathematics, the need for precision is paramount. As a result, when mathematicians use language in doing mathematics, they rely on the literal meaning. Which means, of course, they have to be aware of the literal meaning of what they write or say.

This is why beginning students of mathematics in college are generally offered a crash course in the precise use of language. This may sound like a huge undertaking, given the enormous richness and breadth of everyday language. But the language used in mathematics is so constrained that the task actually turns out to be relatively small. The only thing that makes it difficult is that the student has to

10

learn to eliminate the sloppiness of expression that we are familiar with in everyday life, and instead master a highly constrained, precise (and somewhat stylized) way of writing and speaking.

## 2.1  Mathematical statements

Modern, pure mathematics is primarily concerned with *statements* about *mathematical objects*.

Mathematical objects are things such as: integers, real numbers, sets, functions, etc. Examples of mathematical statements are:

(1)  There are infinitely many prime numbers.

(2)  For every real number $a$, the equation $x^2 + a = 0$ has a real root.

(3)  $\sqrt{2}$ is irrational.

(4)  If $p(n)$ denotes the number of primes less than or equal to the natural number $n$, then as $n$ becomes very large, $p(n)$ approaches $n/\log_e n$.

Not only are mathematicians interested in statements of the above kind, they are, above all, interested in knowing which statements are true and which are false. For instance, in the above examples, (1), (3), and (4) are true whereas (2) is false. The truth or falsity in each case is demonstrated not by observation, measurement, or experiment, as in the sciences, but by a *proof*, about which I will write more in due course.

The truth of (1) can be proved by an ingenious argument known to Euclid.[1] The idea is to show that if we list the primes in increasing order as

$$p_1, p_2, p_3, \ldots, p_n, \ldots$$

then the list must continue for ever. (The first few members of the sequence are: $p_1 = 2$, $p_2 = 3$, $p_3 = 5$, $p_4 = 7$, $p_5 = 11$, $\ldots$)

Consider the list up to some stage $n$:

$$p_1, p_2, p_3, \ldots, p_n$$

The goal is to show that there is another prime that can be added to the list. Provided we do this without assigning $n$ a specific value, this will imply at once that the list is infinite.

Let $N$ be the number we get when we multiply together all the primes we have listed so far and then add 1, i.e.,

$$N = (p_1 \cdot p_2 \cdot p_3 \cdot \ \ldots \ \cdot p_n) + 1$$

---

[1]This proof uses basic facts about prime numbers that will be introduced in Chapter 4, but most readers are likely to be familiar with what is required.

Obviously, $N$ is bigger than all the primes in our list, so if $N$ is prime, we know there is a prime bigger than $p_n$, and hence the list can be continued. (We are not saying that $N$ is that next prime. In fact, $N$ will be much bigger than $p_n$, so it is unlikely to be the next prime.)

Now let's see what happens if $N$ is not prime. Then there must be a prime $q < N$ such that $q$ divides $N$. But none of $p_1, \ldots, p_n$ divides $N$, since the division of $N$ by any one of these leaves a remainder of 1. So, $q$ must be bigger than $p_n$. Thus, once again we see that there is a prime bigger than $p_n$, so the list can be continued.

Since the above argument does not depend in any way upon the value of $n$, it follows that there are infinitely many primes.

Example (2) can easily be proved to be false. Since the square of no real number is negative, the equation $x^2 + 1 = 0$ does not have a real root. Because there is at least one value of $a$ for which the equation $x^2 + a = 0$ does not have a real root (namely $a = 1$), we can conclude that statement (2) is false.

I'll give a proof of (3) later. The only known proofs of (4) are extremely complicated—far too complicated to be included in an introductory text such as this.

Clearly, before we can prove whether a certain statement is true or false, we must be able to understand precisely what the statement says. Above all, mathematics is a very *precise* subject, where exactness of expression is required. This already creates a difficulty, for words tend to be ambiguous, and in real life our use of language is rarely precise.

In particular, when we use language in an everyday setting, we often rely on context to determine what our words convey. An American can truthfully say "July is a summer month," but that would be false if spoken by an Australian. The word "summer" means the same in both statements (namely the hottest three months of the year), but it refers to one part of the year in America and another in Australia.

To take another example, in the phrase "small rodent" the word "small" means something different (in terms of size) than it does in the phrase "small elephant." Most people would agree that a small rodent is a small animal, but a small elephant is definitely not a small animal. The size range referred to by the word "small" can vary depending on the entity to which it is applied.

In everyday life, we use context and our general knowledge of the world and of our lives to fill in missing information in what is written or said, and to eliminate the false interpretations that can result from ambiguities.

For example, we would need to know something about the context in order to correctly understand the statement

*The man saw the woman with a telescope.*

Who had the telescope, the man or the woman?

Ambiguities in newspaper headlines—which are generally written in great haste —can sometimes result in unintended but amusing second readings. Among my favorites that have appeared over the years are:

- Sisters reunited after ten years in checkout line at Safeway.

- Prostitutes appeal to the Pope.

- Large hole appears in High Street. City authorities are looking into it.

- Mayor says bus passengers should be belted.

To systematically make the English language precise (by defining *exactly* what each word is to mean) would be an impossible task. It would also be unnecessary, since people generally do just fine by relying on context and background knowledge.

But in mathematics, things are different. Precision is crucial, and it cannot be assumed that all parties have the same contextual and background knowledge in order to remove ambiguities. Moreover, since mathematical results are regularly used in science and engineering, the cost of miscommunication through an ambiguity can be high, possibly fatal.

At first, it might seem like a herculean task to make the use of language in mathematics sufficiently precise. Fortunately, it is possible because of the special, highly restricted nature of mathematical statements. Every key statement of mathematics (the axioms, conjectures, hypotheses, and theorems) is a positive or negative version of one of four linguistic forms:

(1) Object $a$ has property $P$

(2) Every object of type $T$ has property $P$

(3) There is an object of type $T$ having property $P$

(4) If STATEMENT A, then STATEMENT B

or else is a simple combination of sub-statements of these forms using the connecting words (*combinators*) *and*, *or*, and *not*.

For example,

(1) 3 is a prime number. / 10 is not a prime number.

(2) Every polynomial equation has a complex root. / It is not the case that every polynomial equation has a real root.

(3) There is a prime number between 20 and 25. / There is no even number beyond 2 that is prime.

(4) If $p$ is a prime of the form $4n + 1$, then $p$ is a sum of two squares.

The final statement, about the primes of the form $4n + 1$, is a celebrated theorem of Gauss.

In their everyday work, mathematicians often use more fluent variants of these forms, such as "Not every polynomial equation has a real root" or "No even number is prime except for 2." But those are just that: variants.

The ancient Greek mathematicians seem to have been the first to notice that all mathematical statements can be expressed using one of these simple forms, and they made a systematic study of the language involved, namely the terms *and, or, not, implies, for all,* and *there exists.* They provided universally accepted meanings of these key terms and analyzed their behavior. Today this formal mathematical study is known as *formal logic* or *mathematical logic.*

Mathematical logic is a well established branch of mathematics, studied and used in university departments of mathematics, computer science, philosophy, and linguistics. (It gets a lot more complicated than the original work carried out in ancient Greece by Aristotle and his followers and by the Stoic logicians.)

Some mathematics transition courses and course textbooks include a brief tour through the more basic parts of mathematical logic (as did I in my book *Sets, Functions, and Logic*). But that is not necessary in order to become adept at mathematical thinking. (Many professional mathematicians have virtually no knowledge of mathematical logic.) Consequently, in this book I shall follow a less formal—but still rigorous—path.

## Exercises 2.1.1

1. How would you show that not every number of the form $N = (p_1 \cdot p_2 \cdot p_3 \cdot \ldots \cdot p_n) + 1$ is prime, where $p_1, p_2, p_3, \ldots, p_n, \ldots$ is the list of all prime numbers?

2. Find two unambiguous (but natural sounding) sentences equivalent to the sentence *The man saw the woman with a telescope,* the first where the man has the telescope, the second where the woman has the telescope.

3. For each of the four ambiguous newspaper headlines I stated earlier, rewrite it in a way that avoids the amusing second meaning, while retaining the brevity of a typical headline:

   (a) Sisters reunited after ten years in checkout line at Safeway.

   (b) Prostitutes appeal to the Pope.

   (c) Large hole appears in High Street. City authorities are looking into it.

   (d) Mayor says bus passengers should be belted.

4. The following notice was posted on the wall of a hospital emergency room:

   NO HEAD INJURY IS TOO TRIVIAL TO IGNORE.

   Reformulate to avoid the unintended second reading. (The context for this sentence is so strong that many people have difficulty seeing there is an alternative meaning.)

5. You often see the following notice posted in elevators:

   IN CASE OF FIRE, DO NOT USE ELEVATOR.

This one always amuses me. Comment on the two meanings and reformulate to avoid the unintended second reading. (Again, given the context for this notice, the ambiguity is not problematic.)

6. Official documents often contain one or more pages that are empty apart from one sentence at the bottom:

   *This page intentionally left blank.*

   Does the sentence make a true statement? What is the purpose of making such a statement? What reformulation of the sentence would avoid any logical problems about truth? (Once again, the context means that in practice everyone understands the intended meaning and there is no problem. But the formulation of a similar sentence in mathematics at the start of the twentieth century destroyed one prominent mathematician's seminal work and led to a major revolution in an entire branch of mathematics.)

7. Find (and provide citations for) three examples of published sentences whose literal meaning is (clearly) not what the writer intended. [This is much easier than you might think. Ambiguity is very common.]

8. Comment on the sentence "The temperature is hot today." You hear people say things like this all the time, and everyone understands what is meant. But using language in this sloppy way in mathematics would be disastrous.

9. Provide a context and a sentence within that context, where the word *and* occurs five times in succession, with no other word between those five occurrences. (You are allowed to use punctuation.)

10. Provide a context and a sentence within that context, where the words *and, or, and, or, and* occur in that order, with no other word between them. (Again, you can use punctuation.)

## 2.2   The logical combinators *and*, *or*, and *not*

As a first step in becoming more precise about our use of language (in mathematical contexts), we shall develop precise, unambiguous definitions of the key connecting words *and*, *or*, and *not*. (The other terms, *implies*, *equivalent*, *for all*, and *there exist*, are more tricky and we'll handle them later.)

### The combinator *and*

We need to be able to combine two claims into one that asserts both. For instance, we may wish to say that $\pi$ is greater than 3 *and* less than 3.2. So the word *and* is indispensable.

Sometimes, in order to achieve a completely symbolic expression, we introduce an abbreviation for *and*. The most common ones are

$$\wedge \quad , \quad \&$$

In this book I'll use the former. Thus, the expression

$$(\pi > 3) \wedge (\pi < 3.2)$$

says:

$$\pi \text{ is greater than } 3 \text{ } and \text{ } \pi \text{ is less than } 3.2.$$

In other words, $\pi$ lies between 3 and 3.2 .

There is no possible source of confusion when we use the word *and*. If $\phi$ and $\psi$ are any two mathematical statements,

$$\phi \wedge \psi$$

is the joint assertion (which may or may not be a valid assertion) of *both $\phi$ and $\psi$*. The symbol $\wedge$ is called *wedge*, but the expression $\phi \wedge \psi$ is usually read as "*$\phi$ and $\psi$.*"

The joint statement $\phi \wedge \psi$ (or $\phi \& \psi$) is called the *conjunction* of $\phi$ and $\psi$, and $\phi$, $\psi$ are called the *conjuncts* of the combined statement.[2]

Notice that if $\phi$ and $\psi$ are both true, then $\phi \wedge \psi$ will be true. But if one or both of $\phi$, $\psi$ is false, then so is $\phi \wedge \psi$. In other words, both conjuncts have to be true in order for the conjunction to be true. It only takes one conjunct to be false in order to render the conjunction false.

One thing to notice is that *and* is independent of order in mathematics: $\phi \wedge \psi$ means the same thing as $\psi \wedge \phi$. This is not always true when we use *and* in everyday life. For example,

*John took the free kick and the ball went into the net*

does not mean the same thing as

*The ball went into the net and John took the free kick.*

Mathematicians sometimes use special notation to represent a conjunction of two statements. For example, in dealing with real numbers, we usually write, say,

$$a < x \leq b$$

instead of

$$(a < x) \wedge (x \leq b).$$

---

[2]Introducing formally defined terms to discuss words and concepts we introduce in mathematics, as here, is a common practice. It is impossible to introduce precision without being able to use agreed upon terminology. Likewise, legal contracts frequently have a whole section where the meanings of various terms are stipulated.

*Exercises 2.2.1*

1. The mathematical concept of conjunction captures the meaning of "and" in everyday language. True or false? Explain your answer.

2. Simplify the following symbolic statements as much as you can, leaving your answer in the standard symbolic form. (In case you are not familiar with the notation, I'll answer the first one for you.)

   (a) $(\pi > 0) \wedge (\pi < 10)$   [Answer: $0 < \pi < 10$.]
   (b) $(p \geq 7) \wedge (p < 12)$
   (c) $(x > 5) \wedge (x < 7)$
   (d) $(x < 4) \wedge (x < 6)$
   (e) $(y < 4) \wedge (y^2 < 9)$
   (f) $(x \geq 0) \wedge (x \leq 0)$

3. Express each of your simplified statements from Question 1 in natural English.

4. What strategy would you adopt to show that the conjunction $\phi_1 \wedge \phi_2 \wedge \ldots \wedge \phi_n$ is true?

5. What strategy would you adopt to show that the conjunction $\phi_1 \wedge \phi_2 \wedge \ldots \wedge \phi_n$ is false?

6. Is it possible for one of $(\phi \wedge \psi) \wedge \theta$ and $\phi \wedge (\psi \wedge \theta)$ to be true and the other false, or does the associative property hold for conjunction? Prove your answer.

7. Which of the following is more likely?

   (a) Alice is a rock star and works in a bank.
   (b) Alice is quiet and works in a bank.
   (c) Alice is quiet and reserved and works in a bank.
   (d) Alice is honest and works in a bank.
   (e) Alice works in a bank.

   If you believe there is no definite answer, say so.

8. In the following table, T denotes 'true' and F denotes 'false'. The first two columns list all the possible combinations of values of T and F that the two statements $\phi$ and $\psi$ can have. The third column should give the truth value (T or F) $\phi \wedge \psi$ achieves according to each assignment of T or F to $\phi$ and $\psi$.

| $\phi$ | $\psi$ | $\phi \wedge \psi$ |
|---|---|---|
| T | T | ? |
| T | F | ? |
| F | T | ? |
| F | F | ? |

Fill in the final column. The resulting table is an example of a "propositional truth table."

## The combinator *or*

We wish to be able to assert that statement $A$ is true *or* statement $B$ is true. For instance, we might want to say

$$a > 0 \text{ } or \text{ the equation } x^2 + a = 0 \text{ has a real root}$$

or perhaps we want to say

$$ab = 0 \text{ if } a = 0 \text{ } or \text{ } b = 0$$

These two simple examples show we face a potential ambiguity. The meaning of *or* is different in these two cases. In the first assertion there is no possibility of both eventualities occurring at once. Moreover, the meaning is unchanged if we put the word *either* at the beginning of the sentence. In the second case, it is quite possible for *both* $a$ and $b$ to be zero.[3]

But mathematics has no place for potential ambiguity in the meaning of such a commonly used word as *or*, so we must choose one or the other meaning. It turns out to be more convenient to opt for the inclusive use. Accordingly, whenever we use the word *or* in mathematics we always mean the inclusive or. If $\phi$ and $\psi$ are mathematical statements, $\phi$ *or* $\psi$ asserts that at least one of $\phi, \psi$ is valid. We use the symbol

$$\vee$$

to denote (*inclusive*) *or*. Thus

$$\phi \vee \psi$$

means that *at least one of* $\phi, \psi$ is valid. The symbol $\vee$ is called *vee*, but mathematicians usually read $\phi \vee \psi$ as "$\phi$ *or* $\psi$."

We call $\phi \vee \psi$ the *disjunction* of $\phi$ and $\psi$, and refer to each of $\phi, \psi$ as the *disjuncts* of the combined statement.

It requires only one of $\phi, \psi$ to be true in order for the disjunction $\phi \vee \psi$ to be true.

For instance, the following (rather silly) statement is true:

$$(3 < 5) \vee (1 = 0)$$

Although this is a silly example, you should pause for a moment and make sure you understand why this statement is not only mathematically *meaningful* but actually true. Silly examples are often useful to help understand tricky concepts—and disjunction can be tricky.

---

[3] Even if we were to alter our second assertion by inserting the word *either*, we would still read it as allowing for both possibilities to occur at once, since the use of the word *either* only serves to strengthen the exclusivity in an assertion *where it is already clear* that there is no possibility of both, as in the first assertion.

*Exercises 2.2.2*

1. Simplify the following symbolic statements as much as you can, leaving your answer in a standard symbolic form (assuming you are familiar with the notation):

    (a) $(\pi > 3) \lor (\pi > 10)$

    (b) $(x < 0) \lor (x > 0)$

    (c) $(x = 0) \lor (x > 0)$

    (d) $(x > 0) \lor (x \geq 0)$

    (e) $(x > 3) \lor (x^2 > 9)$

2. Express each of your simplified statements from Question 1 in natural English.

3. What strategy would you adopt to show that the disjunction $\phi_1 \lor \phi_2 \lor \ldots \lor \phi_n$ is true?

4. What strategy would you adopt to show that the disjunction $\phi_1 \lor \phi_2 \lor \ldots \lor \phi_n$ is false?

5. Is it possible for one of $(\phi \lor \psi) \lor \theta$ and $\phi \lor (\psi \lor \theta)$ to be true and the other false, or does the associative property hold for disjunction? Prove your answer.

6. Which of the following is more likely?

    (a) Alice is a rock star or she works in a bank.

    (b) Alice is quiet and works in a bank.

    (c) Alice is a rock star.

    (d) Alice is honest and works in a bank.

    (e) Alice works in a bank.

    If you believe there is no definite answer, say so.

7. Fill in the entries in the final column of the following truth table:

| $\phi$ | $\psi$ | $\phi \lor \psi$ |
|---|---|---|
| T | T | ? |
| T | F | ? |
| F | T | ? |
| F | F | ? |

**The combinator *not***

Many mathematical statements involve a negation, i.e., a claim that a particular statement is false.

If $\psi$ is any statement,

$$not\text{-}\psi$$

is the statement that $\psi$ is false. It is called the *negation* of $\psi$.

Thus, if $\psi$ is a true statement, $not\text{-}\psi$ is a false statement, and if $\psi$ is a false statement, $not\text{-}\psi$ is a true statement. Nowadays, the most commonly used symbolic abbreviation for $not\text{-}\psi$ is

$$\neg\psi$$

but older texts sometimes use $\sim\psi$.

In certain circumstances we use special notations for negation. For instance, we generally use the more familiar

$$x \neq y$$

instead of

$$\neg(x = y)$$

On the other hand, we would probably write

$$\neg(a < x \leq b)$$

instead of

$$a \not< x \not\leq b$$

as the latter is ambiguous. (We could make it precise, but it seems rather inelegant, and mathematicians don't do it.)

Although the mathematical usage of the word *not* accords with most common usage, negation is sometimes used very loosely in everyday speech, so you have to be careful. For instance, there is no confusion about the meaning of the statement

$$\neg(\pi < 3).$$

This clearly means

$$\pi \geq 3$$

which, incidentally, is the same as

$$(\pi = 3) \vee (\pi > 3).$$

But consider the statement

*All foreign cars are badly made.*

What is the negation of this statement? For instance, is it any one of the following?

(a) *All foreign cars are well made.*

(b) *All foreign cars are not badly made.*

(c) *At least one foreign car is well made.*

(d) *At least one foreign car is not badly made.*

A common mistake is for the beginner to choose (a). But this is easily seen to be wrong. The original statement is surely false. Hence the *negation* of that statement will be true. But (a) is certainly not true! Neither is (b) true. So realistic considerations lead us to conclude that if the correct answer is to be found in the above list, then it has to be either (c) or (d). (We shall later see how we can eliminate (a) and (b) by a formal mathematical argument.)

In point of fact, both (c) and (d) can be said to represent the negation of our original statement. (Any well made foreign car testifies the *truth* of both (c) and (d).) Which do you think most closely corresponds to the negation of the original statement?

We shall return to this example later, but before we leave it for now, let us note that the original statement is only concerned with foreign cars. Hence its negation will only deal with foreign cars. So, the negation will not involve any reference to *domestic* cars. For instance, the statement

*All domestic cars are well made*

cannot be the negation of our original statement. Indeed, knowing whether our original statement is true or false in no way helps us to decide the truth or falsity of the above statement. To be sure, *domestic* is the negation of *foreign* in this context, but we are negating the assertion as a whole, not some adjective occurring in it.

Now you might appreciate why it is important to analyze the use of language before we use it in mathematics. In the case of examples about cars, we can use our knowledge of the world to sort out what is true and what is false. But when it comes to mathematics, we often do not have enough background knowledge. The statements we write down may constitute all we know.

## Exercises 2.2.3

1. Simplify the following symbolic statements as much as you can, leaving your answer in a standard symbolic form (assuming you are familiar with the notation):

   (a) $\neg(\pi > 3.2)$

   (b) $\neg(x < 0)$

   (c) $\neg(x^2 > 0)$

   (d) $\neg(x = 1)$

   (e) $\neg\neg\psi$

2. Express each of your simplified statements from Question 1 in natural English.

3. Is showing that the negation $\neg\phi$ is true the same as showing that $\phi$ is false? Explain your answer.

4. Fill in the entries in the final column of the following truth table:

| $\phi$ | $\neg\phi$ |
|---|---|
| T | ? |
| F | ? |

5. Let $D$ be the statement "The dollar is strong", $Y$ the statement "The Yuan is strong," and $T$ the statement "New US–China trade agreement signed". Express the main content of each of the following (fictitious) newspaper headlines in logical notation. (Note that logical notation captures truth, but not the many nuances and inferences of natural language.) Be prepared to justify and defend your answers.

   (a) Dollar and Yuan both strong

   (b) Trade agreement fails on news of weak Dollar

   (c) Dollar weak but Yuan strong, following new trade agreement

   (d) Strong Dollar means a weak Yuan

   (e) Yuan weak despite new trade agreement, but Dollar remains strong

   (f) Dollar and Yuan can't both be strong at same time.

   (g) If new trade agreement is signed, Dollar and Yuan can't both remain strong

   (h) New trade agreement does not prevent fall in Dollar and Yuan

   (i) US–China trade agreement fails but both currencies remain strong

   (j) New trade agreement will be good for one side, but no one knows which.

6. In US law, a trial verdict of "Not guilty" is given when the prosecution fails to prove guilt. This, of course, does not mean the defendant is, as a matter of actual fact, innocent. Is this state of affairs captured accurately when we use "not" in the mathematical sense? (i.e., Do "Not guilty" and "$\neg$ guilty" mean the same thing?) What if we change the question to ask if "Not proven" and "$\neg$ proven" mean the same thing?

7. The truth table for $\neg\neg\phi$ is clearly the same as that for $\phi$ itself, so the two expressions make identical truth assertions. This is not necessarily true for negation in everyday life. For example, you might find yourself saying "I was not displeased with the movie." In terms of formal negation, this has the form $\neg(\neg\,\text{PLEASED})$, but your statement clearly does not mean that you were pleased with the movie. Indeed, it means something considerably less positive. How would you capture this kind of use of language in the formal framework we have been looking at?

## 2.3   Implication

Now things get really tricky. Brace yourself for several days of confusion until the ideas sort themselves out in your mind.

In mathematics we frequently encounter expressions of the form

$$(*) \quad \phi \; implies \; \psi$$

Indeed implication provides the means by which we prove statements, starting from initial observations or axioms. The question is, what is the meaning of an assertion of the form $(*)$?

It would not be unreasonable to assume it means the following:

*If* $\phi$ is true, *then* $\psi$ has to be true as well.

But the carefully crafted, lawyer-like wording I used to introduce that possible meaning should indicate that things are slippery.

Suppose that for $\phi$ we take the true assertion '$\sqrt{2}$ is irrational' (we'll prove that later on) and for $\psi$ we take the true assertion '$0 < 1$'. Then is the expression $(*)$ true? In other words, does the irrationality of $\sqrt{2}$ imply that 0 is less than 1? Of course it does not. There is no meaningful connection between the two statements $\phi$ and $\psi$ in this case.

The point is, *implies* entails causality. This was not a consideration in the case of *and* and *or*. There is nothing to stop us conjoining or disjoining two totally unrelated statements. For example, there is no difficulty in determining the truth of the statements

$$(\text{Julius Caesar is dead}) \wedge (1 + 1 = 2)$$

$$(\text{Julius Caesar is dead}) \vee (1 + 1 = 2)$$

(Once again, I am using a frivolous example to illustrate a tricky point. Since mathematics is frequently applied to real-world situations, we may well encounter a statement that combines the two domains, mathematics and the real world.)

Thus, in adopting precise meanings for the words *and*, *or*, and *not*, we were able to ignore the meanings of the component statements and focus entirely on their truth values (i.e., are the statements true or false?).

In the process, we did of course have to make choices which gave some of the terms meanings that were different from their everyday language counterparts. We had to stipulate that *or* means inclusive-or and adopt a minimalistic interpretation of *not* reminiscent of the "not proven" verdict in a court of law.

We will have to adopt a similar approach to *implies* in order to come up with an unambiguous meaning that depends only on truth and falsity. But in this case we have to be far more aggressive—so much so, that to avoid any possible confusion, we will have to use a different term than "implies".

As I noted once already, the problem is that when we say "$\phi$ implies $\psi$", we mean that $\phi$ somehow causes or brings about $\psi$. This entails that the truth of $\psi$

follows from the truth of $\phi$, but *truth alone does not fully capture what is meant by the word "implies"*. Not even close. So we had better agree to not use the word *implies* unless we mean it.

The approach we shall adopt is to separate the notion of implication into two parts, the truth part and the causation part. The truth part is generally known as the *conditional*, or sometimes the *material conditional*. Thus we have the relationship:

$$\text{implication} = \text{conditional} + \text{causation}$$

We will use the symbol $\Rightarrow$ to denote the conditional operator. Thus,

$$\phi \Rightarrow \psi$$

denotes the truth part of $\phi$ *implies* $\psi$.

(Modern mathematical logic texts generally use a single arrow, $\rightarrow$, instead of $\Rightarrow$, but to avoid confusion with the notation for functions you are likely to meet later in your mathematical education, I'll use the more old-fashioned, double-arrow notation for the conditional.)

Any expression of the form

$$\phi \Rightarrow \psi$$

is referred to as *a conditional expression*, or simply *a conditional*. We refer to $\phi$ as the *antecedent* of the conditional and $\psi$ as the *consequent*.

The truth or falsity of a conditional will be *defined* entirely in terms of the truth or falsity of the antecedent and the consequent. That is to say, whether or not the conditional expression $\phi \Rightarrow \psi$ is true will depend entirely upon the truth or falsity of $\phi$ and $\psi$, taking no account of whether or not there is any meaningful connection between $\phi$ and $\psi$.

The reason this approach turns out to be a useful one, is that in all cases where there *is* a meaningful and genuine implication $\phi$ *implies* $\psi$, the conditional $\phi \Rightarrow \psi$ does accord with that implication.

In other words, it will turn out that our defined notion $\phi \Rightarrow \psi$ fully captures $\phi$ *implies* $\psi$, whenever there is a genuine implication. But our notion extends beyond that to cover all cases where we know the truth and falsity of $\phi$ and $\psi$ but there is no meaningful connection between the two.

Since we are ignoring causation, which is a highly significant aspect of the notion of implication, our definition may (and in fact will) turn out to have consequences that are at the very least counterintuitive, and possibly even absurd. But those will be restricted to situations where there is no genuine implication.

The task we face, then, is to stipulate rules that will enable us to complete the truth table

| $\phi$ | $\psi$ | $\phi \Rightarrow \psi$ |
|--------|--------|-------------------------|
| T | T | ? |
| T | F | ? |
| F | T | ? |
| F | F | ? |

The first rule is easy. If there is a valid, genuine implication $\phi$ *implies* $\psi$, then the truth of $\phi$ implies the truth of $\psi$. So the first row of the table has to have T everywhere:

| $\phi$ | $\psi$ | $\phi \Rightarrow \psi$ |
|---|---|---|
| T | T | T |
| T | F | ? |
| F | T | ? |
| F | F | ? |

*Exercises 2.3.1*

1. Fill in the second row of the truth table.

2. Provide a justification of your entry.

Before I complete the second row of the truth table (and thereby tell you the answers to the above exercises—so you should do them before reading on), let's take a look at the consequences of the choice we made in completing the first row the way we did.

If we know that the statement $N > 7$ is true, then we can conclude that $N^2 > 40$ is true. According to the first row of our table,

$$(N > 7) \Rightarrow (N^2 > 40)$$

is true. This is entirely consistent with the validity of the genuine (causal) implication: $N > 7$ implies $N^2 > 40$.

But what happens if $\phi$ is the true statement "Julius Caesar is dead" and $\psi$ is the true statement "$\pi > 3$"? According to the first row of our table, the conditional

$$(\text{Julius Caesar is dead}) \Rightarrow (\pi > 3)$$

has the value T.

In real-world terms, there is of course no relationship between the true fact that Julius Caesar is dead and the true fact that $\pi$ is greater than 3. But so what? The conditional does not claim to capture causality relationships, or indeed meaningful relationships of any kind. The truth of $[(\text{Julius Caesar is dead}) \Rightarrow (\pi > 3)]$ only becomes problematic if you interpret the conditional ($\Rightarrow$) as implication. The cost of defining $[\phi \Rightarrow \psi]$ so it always has a well-defined truth value (which is a mathematically valuable property) is that we have to get used to not reading more into the conditional than is given by the definition.

Now let's continue with the task of filling in the truth table for the conditional. If $\phi$ is true and $\psi$ is false, then there is no way that $\phi$ can genuinely imply $\psi$. (Why? Well, if there were a genuine implication, then the truth of $\psi$ would follow *automatically* from the truth of $\phi$.) So if $\phi$ is true and $\psi$ is false, the genuine implication must be false. Hence the conditional $[\phi \Rightarrow \psi]$ should be false as well, and the table now looks like this:

| $\phi$ | $\psi$ | $\phi \Rightarrow \psi$ |
|:---:|:---:|:---:|
| T | T | T |
| T | F | F |
| F | T | ? |
| F | F | ? |

*Exercises 2.3.2*

1. Fill in the third and fourth rows of the truth table.

2. Provide justifications for your entries.

(I'll get to the third and fourth rows in a moment, so you should do the above exercise before you read on.)

At this point, you should probably go back to the start of the discussion of *implies* and re-read everything we have done so far. Though it might seem we are making much ado about nothing, this entire discussion is typical of work to provide precise definitions of the fundamental concepts of mathematics.

Though the use of simple (and often silly) examples can give the impression of it all being an irrelevant game, the consequences are far from irrelevant. The next time you step into an airplane, be aware that the flight control software on which your life depends makes use of the formal notions of $\wedge, \vee, \neg$, and $\Rightarrow$ we are discussing here. And part of what it takes to make that software reliable is that the system never encounters a mathematical statement whose truth is not defined. You, as a human being, only care about statements of the form $[\phi \Rightarrow \psi]$ when everything makes sense. But computer systems do not have a notion of "making sense." They deal in the binary logic of truth and falsity. What matters to a computer system is that everything is always precisely defined, with a specific truth value.

Once we get used to ignoring all questions of causality, the truth-values of the conditional seem straightforward enough when the antecedent is true. (If they don't, you should go back and read that discussion yet again. There was a reason I suggested you do that!) But what about the last two rows in the truth table, where the antecedent is false?

To deal with this case, we consider not the notion of implication, but its negation. We extract the causation-free, truth-value part of the statement "$\phi$ does not imply $\psi$", which I shall write as

$$\phi \not\Rightarrow \psi$$

Leaving aside all question of whether there is a meaningful causal relation between $\phi$ and $\psi$ and concentrating solely on truth values, how can we be sure that "$\phi$ does not imply $\psi$" is a valid statement? More precisely, how should the truth or falsity of the statement $\phi \not\Rightarrow \psi$ depend upon the truth or falsity of $\phi$ and $\psi$?

Well, in terms of truth values, $\phi$ will *not* imply $\psi$ if it is the case that *although* $\phi$ is true, $\psi$ is *nevertheless* false.

Please read that last sentence again. Now once more. Okay, now we are ready to continue.[4]

We therefore define $\phi \nRightarrow \psi$ to be true precisely in case $\phi$ is true and $\psi$ is false.

Having defined the truth or falsity of $\phi \nRightarrow \psi$, we obtain that of $\phi \Rightarrow \psi$ by just taking the negation. The conditional $\phi \Rightarrow \psi$ will be true exactly when $\phi \nRightarrow \psi$ is false.

Examination of this definition leads to the conclusion: $\phi \Rightarrow \psi$ will be true whenever one of the following holds:

(1) $\phi$ and $\psi$ are both true.

(2) $\phi$ is false and $\psi$ is true.

(3) $\phi$ and $\psi$ are both false.

The complete truth table thus looks like this:

| $\phi$ | $\psi$ | $\phi \Rightarrow \psi$ |
|:---:|:---:|:---:|
| T | T | T |
| T | F | F |
| F | T | T |
| F | F | T |

The points to note about this are:

(a) We are *defining* a notion (the conditional) that captures only part of what 'implies' means.

(b) To avoid difficulties, we base our definition solely on the notion of truth and falsity.

(c) Our definition agrees with our intuition concerning implication in all meaningful cases.

(d) The definition for a true antecedent is based on an analysis of the truth-values of genuine implication.

(e) The definition for a false antecedent is based on a truth-value analysis of the notion that $\phi$ does not imply $\psi$.

Summing up, in defining the conditional the way we do, we do not end up with a notion that *contradicts* the notion of (genuine) implication. Rather, we obtain a notion that *extends* (genuine) implication to cover those cases where a claim of implication is irrelevant (the antecedent is false) or meaningless (there is no real connection between the antecedent and the consequent). In the meaningful case where there is a relationship between $\phi$ and $\psi$ and in addition $\phi$ is true, namely

---

[4]On second thought, maybe you should read it a fourth time just to be sure.

the cases covered by the first two rows of the table, the truth value of the conditional will be the same as the truth value of the actual implication.

Remember, it is the fact that the conditional always has a well-defined truth value that makes this notion important in mathematics, since in mathematics (as well as in aircraft control systems!) we cannot afford to have statements with undefined truth values floating around.

*Exercises 2.3.3*

1. Which of the following are true and which are false?

    (a) $(\pi^2 > 2) \Rightarrow (\pi > 1.4)$

    (b) $(\pi^2 < 0) \Rightarrow (\pi = 3)$

    (c) $(\pi^2 > 0) \Rightarrow (1 + 2 = 4)$

    (d) $(\pi < \pi^2) \Rightarrow (\pi = 5)$

    (e) $(e^2 \geq 0) \Rightarrow (e < 0)$

    (f) $\neg(5 \text{ is an integer}) \Rightarrow (5^2 \geq 1)$

    (g) (The area of a circle of radius 1 is $\pi$) $\Rightarrow$ (3 is prime)

    (h) (Squares have three sides) $\Rightarrow$ (Triangles have four sides)

    (i) (Elephants can climb trees) $\Rightarrow$ (3 is irrational)

    (j) (Euclid's birthday was July 4) $\Rightarrow$ (Rectangles have four sides)

2. As in Exercise 2.2.3(5), let $D$ be the statement "The dollar is strong," $Y$ the statement "The Yuan is strong," and $T$ the statement "New US–China trade agreement signed." Express the main content of each of the following (fictitious) newspaper headlines in logical notation. (Remember, logical notation captures truth, but not the many nuances and inferences of natural language.) As before, be prepared to justify and defend your answers.

    (a) New trade agreement will lead to strong currencies in both countries.

    (b) If the trade agreement is signed, a rise in the Yuan will result in a fall in the Dollar.

    (c) Dollar weak but Yuan strong, following new trade agreement

    (d) Strong Dollar means a weak Yuan

    (e) New trade agreement means Dollar and Yuan will be tightly linked.

3. Complete the following truth table

| $\phi$ | $\neg\phi$ | $\psi$ | $\phi \Rightarrow \psi$ | $\neg\phi \vee \psi$ |
|---|---|---|---|---|
| T | ? | T | ? | ? |
| T | ? | F | ? | ? |
| F | ? | T | ? | ? |
| F | ? | F | ? | ? |

Note: $\neg$ has the same binding rules as $-$ (minus) in arithmetic and algebra, so $\neg\phi \lor \psi$ is the same as $(\neg\phi) \lor \psi$.

4. What conclusions can you draw from the above table?

5. Complete the following truth table. (Recall that $\phi \not\Rightarrow \psi$ is another way of writing $\neg[\phi \Rightarrow \psi]$.)

| $\phi$ | $\psi$ | $\neg\psi$ | $\phi \Rightarrow \psi$ | $\phi \not\Rightarrow \psi$ | $\phi \land \neg\psi$ |
|---|---|---|---|---|---|
| T | T | ? | ? | ? | ? |
| T | F | ? | ? | ? | ? |
| F | T | ? | ? | ? | ? |
| F | F | ? | ? | ? | ? |

6. What conclusions can you draw from the above table?

Closely related to implication is the notion of *equivalence*. Two statements $\phi$ and $\psi$ are said to be *(logically) equivalent* if each implies the other. The analogous, formal notion defined in terms of the conditional is known as the *biconditional*. We shall write the biconditional as

$$\phi \Leftrightarrow \psi$$

(Modern logic texts use the notation $\phi \leftrightarrow \psi$.) The biconditional is formally defined to be an abbreviation for the conjunction

$$(\phi \Rightarrow \psi) \land (\psi \Rightarrow \phi)$$

Looking back at the definition of the conditional, this means that the biconditional $\phi \Leftrightarrow \psi$ will be true if $\phi$ and $\psi$ are both true or both false, and $\phi \Leftrightarrow \psi$ will be false if exactly one of $\phi, \psi$ is true and the other false.

One way to show that two logical expressions are biconditionally-equivalent is to show that their truth tables are the same. Consider, for example, the expression $(\phi \land \psi) \lor (\neg\phi)$. We can build its table column by column as follows:

| $\phi$ | $\psi$ | $\phi \land \psi$ | $\neg\phi$ | $(\phi \land \psi) \lor (\neg\phi)$ |
|---|---|---|---|---|
| T | T | T | F | T |
| T | F | F | F | F |
| F | T | F | T | T |
| F | F | F | T | T |

The final column is the same as that for $\phi \Rightarrow \psi$. Hence, $(\phi \land \psi) \lor (\neg\phi)$ is biconditionally equivalent to $\phi \Rightarrow \psi$.

We can also draw up tables for expressions involving more than two basic statements, such as $(\phi \land \psi) \lor \theta$, which has three, but if there are $n$ constituent statements involved there will be $2^n$ rows in the table, so already $(\phi \land \psi) \lor \theta$ needs 8 rows!

*Exercises 2.3.4*

1. Build a truth table to prove the claim I made earlier that $\phi \Leftrightarrow \psi$ is true if $\phi$ and $\psi$ are both true or both false and $\phi \Leftrightarrow \psi$ is false if exactly one of $\phi, \psi$ is true and the other false. (To constitute a proof, your table should have columns that show how the entries for $\phi \Leftrightarrow \psi$ are derived, one operator at a time, as in the previous exercises.)

2. Build a truth table to show that

$$(\phi \Rightarrow \psi) \Leftrightarrow (\neg \phi \vee \psi)$$

is true for all truth values of $\phi$ and $\psi$. A statement whose truth values are all T is called a *logical validity*, or sometimes a *tautology*.

3. Build a truth table to show that

$$(\phi \not\Rightarrow \psi) \Leftrightarrow (\phi \wedge \neg \psi)$$

is a tautology.

4. The ancient Greeks formulated a basic rule of reasoning for proving mathematical statements. Called *modus ponens*, it says that if you know $\phi$ and you know $\phi \Rightarrow \psi$, then you can conclude $\psi$.

    (a) Construct a truth table for the logical statement

    $$[\phi \wedge (\phi \Rightarrow \psi)] \Rightarrow \psi$$

    (b) Explain how the truth table you obtain demonstrates that *modus ponens* is a valid rule of inference.

5. *Mod-2 arithmetic* has just the two numbers 0 and 1 and follows the usual rules of arithmetic together with the additional rule $1 + 1 = 0$. (It is the arithmetic that takes place in a single bit location in a digital computer.) Complete the following table:

| $M$ | $N$ | $M \times N$ | $M + N$ |
|---|---|---|---|
| 1 | 1 | ? | ? |
| 1 | 0 | ? | ? |
| 0 | 1 | ? | ? |
| 0 | 0 | ? | ? |

6. In the table you obtained in the above exercise, interpret 1 as T and 0 as F and view $M, N$ as statements.

    (a) Which of the logical combinators $\wedge$, $\vee$ corresponds to $\times$ ?

(b) Which logical combinator corresponds to $+$ ?

(c) Does $\neg$ correspond to $-$ (minus)?

7. Repeat the above exercise, but interpret 0 as T and 1 as F. What conclusions can you draw?

8. The following puzzle was introduced by the psychologist Peter Wason in 1966, and is one of the most famous subject tests in the psychology of reasoning. Most people get it wrong. (So you have been warned!)

   Four cards are placed on the table in front of you. You are told (truthfully) that each has a letter printed on one side and a digit on the other, but of course you can only see one face of each. What you see is:

$$\text{B} \quad \text{E} \quad 4 \quad 7$$

   You are now told that the cards you are looking at were chosen to follow the rule "If there is a vowel on one side, then there is an odd number on the other side." What is the least number of cards you *have* to turn over to verify this rule, and which cards do you in fact *have* to turn over?

There is some terminology associated with implication (i.e., real implication, not just the conditional) that should be mastered straight away, as it pervades all mathematical discussion.

In an implication

$$\phi \ implies \ \psi$$

we call $\phi$ the *antecedent* and $\psi$ the *consequent*.

The following all mean the same:

(1) $\phi$ implies $\psi$

(2) if $\phi$ then $\psi$

(3) $\phi$ is sufficient for $\psi$

(4) $\phi$ only if $\psi$

(5) $\psi$ if $\phi$

(6) $\psi$ whenever $\phi$

(7) $\psi$ is necessary for $\phi$

The first four all mention $\phi$ before $\psi$, and of these the first three seem obvious enough. But caution is required in the case of (4). Notice the contrast between (4) and (5) as far as the order of $\phi$ and $\psi$ is concerned. Beginners often encounter considerable difficulty in appreciating the distinction between *if* and *only if*.

Likewise, the use of the word *necessary* in (7) often causes confusion. Notice that to say that $\psi$ is a necessary condition for $\phi$ does not mean that $\psi$ on its own is enough to guarantee $\phi$. Rather what it says is that $\psi$ will have to hold *before there can even be any question of $\phi$ holding*. For that to be the case, $\phi$ must imply $\psi$. (This is another of those occasions where my strong advice would be to re-read this paragraph several times until you are sure you get the point—then read it at least one more time!)

The following diagram might help you remember the distinction between 'necessary' and 'sufficient':

(Think of the word 'sun'. This will remind you of the order.)

Because equivalence reduces to implication both ways, it follows from the above discussion that the following are also equivalent:

(1) $\phi$ is equivalent to $\psi$

(2) $\phi$ is necessary and sufficient for $\psi$

(3) $\phi$ if and only if $\psi$

A common abbreviation for the phrase *if and only if* is *iff* (or occasionally *iffi*). Thus we often write

$$\phi \text{ iff } \psi$$

to mean $\phi$ and $\psi$ are equivalent.

Note that if we were to be strict about the matter, the above discussion of equivalent terminologies refers to implication and equivalence, not their formal counterparts the conditional and the biconditional. However, mathematicians frequently use the symbol $\Rightarrow$ as an abbreviation for *implies* and $\Leftrightarrow$ as an abbreviation for *is equivalent to*, so the different terminologies often are used together with these formally defined symbols.

Although this is invariably confusing to beginners, it's simply the way mathematical practice has evolved, so there is no getting around it. You would be entirely justified in throwing your hands up at what seems on the face of it to be sloppy practice. After all, if there are genuine problems with the meanings of certain words that necessitate a lengthy discussion and the formulation of formal definitions of concepts that are not identical with their everyday counterparts (such as the difference between the conditional and implication), why do mathematicians then promptly revert to the everyday notions that they began by observing to be problematic?

Here is why the professionals do this: The conditional and biconditional only differ from implication and equivalence in situations *that do not arise in the course of normal mathematical practice.* In any real mathematical context, the conditional "is" implication and the biconditional "is" equivalence. Thus, having made note of where the formal notions differ from the everyday ones, mathematicians simply move on and turn their attention to other things. (Computer programmers and people who develop aircraft control systems do not have such freedom.)

*Exercises 2.3.5*

1. One way to prove that

$$\neg(\phi \wedge \psi) \quad \text{and} \quad (\neg\phi) \vee (\neg\psi)$$

are equivalent is to show they have the same truth table:

| $\phi$ | $\psi$ | $\phi \wedge \psi$ | $\neg(\phi \wedge \psi)$ <br> * | $\neg\phi$ | $\neg\psi$ | $(\neg\phi) \vee (\neg\psi)$ <br> * |
|---|---|---|---|---|---|---|
| T | T | T | F | F | F | F |
| T | F | F | T | F | T | T |
| F | T | F | T | T | F | T |
| F | F | F | T | T | T | T |

Since the two columns marked * are identical, we know that the two expressions are equivalent.

Thus, negation has the effect that it changes $\vee$ into $\wedge$ and changes $\wedge$ into $\vee$. An alternative way to prove this is to argue directly with the meaning of the first statement:

1. $\phi \wedge \psi$ means both $\phi$ and $\psi$ are true.

2. Thus $\neg(\phi \wedge \psi)$ means it is not the case that both $\phi$ and $\psi$ are true.

3. If they are not both true, then at least one of $\phi$, $\psi$ must be false.

4. This is clearly the same as saying that at least one of $\neg\phi$ and $\neg\psi$ is true. (By the definition of negation).

5. By the meaning of *or*, this can be expressed as $(\neg\phi) \vee (\neg\psi)$.

Provide an analogous logical argument to show that $\neg(\phi \vee \psi)$ and $(\neg\phi) \wedge (\neg\psi)$ are equivalent.

2. By a *denial* of a statement $\phi$ we mean any statement equivalent to $\neg\phi$. Give a useful denial of each of the following statements.

   (a) 34,159 is a prime number.

   (b) Roses are red and violets are blue.

   (c) If there are no hamburgers, I'll have a hot dog.

    (d) Fred will go but he will not play.

    (e) The number $x$ is either negative or greater than 10.

    (f) We will win the first game or the second.

3. Which of the following conditions is *necessary* for the natural number $n$ to be divisible by 6?

    (a) $n$ is divisible by 3.

    (b) $n$ is divisible by 9.

    (c) $n$ is divisible by 12.

    (d) $n = 24$.

    (e) $n^2$ is divisible by 3.

    (f) $n$ is even and divisible by 3.

4. In Exercise 3, which conditions are *sufficient* for $n$ to be divisible by 6?

5. In Exercise 3, which conditions are *necessary and sufficient* for $n$ to be divisible by 6?

6. Let $m, n$ denote any two natural numbers. Prove that $mn$ is odd iff $m$ and $n$ are odd.

7. With reference to the previous question, is it true that $mn$ is even iff $m$ and $n$ are even?

8. Show that $\phi \Leftrightarrow \psi$ is equivalent to $(\neg \phi) \Leftrightarrow (\neg \psi)$. How does this relate to your answers to Questions 6 and 7 above?

9. Construct truth tables to illustrate the following:

    (a) $\phi \Leftrightarrow \psi$

    (b) $\phi \Rightarrow (\psi \vee \theta)$

10. Use truth tables to prove that the following are equivalent:

    (a) $\neg(\phi \Rightarrow \psi)$ and $\phi \wedge (\neg \psi)$

    (b) $\phi \Rightarrow (\psi \wedge \theta)$ and $(\phi \Rightarrow \psi) \wedge (\phi \Rightarrow \theta)$

    (c) $(\phi \vee \psi) \Rightarrow \theta$ and $(\phi \Rightarrow \theta) \wedge (\psi \Rightarrow \theta)$

11. Verify the equivalences in (b) and (c) in the previous question by means of a logical argument. (So, in the case of (b), for example, you must show that assuming $\phi$ and deducing $\psi \wedge \theta$ is the same as both deducing $\psi$ from $\phi$ and $\theta$ from $\phi$.)

12. Use truth tables to prove the equivalence of $\phi \Rightarrow \psi$ and $(\neg\psi) \Rightarrow (\neg\phi)$.

    $(\neg\psi) \Rightarrow (\neg\phi)$ is called the *contrapositive* of $\phi \Rightarrow \psi$. The logical equivalence of a conditional and its contrapositive means that one way to prove an implication is to verify the contrapositive. This is a common form of proof in mathematics that we'll encounter later.

13. Write down the contrapositives of the following statements:

    (a) If two rectangles are congruent, they have the same area.

    (b) If a triangle with sides $a, b, c$ ($c$ largest) is right-angled, then $a^2 + b^2 = c^2$.

    (c) If $2^n - 1$ is prime, then $n$ is prime.

    (d) If the Yuan rises, the Dollar will fall.

14. It is important not to confuse the contrapositive of a conditional $\phi \Rightarrow \psi$ with its *converse* $\psi \Rightarrow \phi$. Use truth tables to show that the contrapositive and the converse of $\phi \Rightarrow \psi$ are not equivalent.

15. Write down the converses of the four statements in Question 13.

16. Show that for any two statements $\phi$ and $\psi$ cither $\phi \Rightarrow \psi$ or its converse is true (or both). This is another reminder that the conditional is not the same as implication.

17. Express the combinator

    $$\phi \text{ unless } \psi$$

    in terms of the standard logical combinators.

18. Identify the antecedent and the consequent in each of the following conditionals:

    (a) If the apples are red, they are ready to eat.

    (b) The differentiability of a function $f$ is sufficient for $f$ to be continuous.

    (c) A function $f$ is bounded if $f$ is integrable.

    (d) A sequence $s$ is bounded whenever $s$ is convergent.

    (e) It is necessary that $n$ is prime in order for $2^n - 1$ to be prime.

    (f) The team wins only when Karl is playing.

    (g) When Karl plays the team wins.

    (h) The team wins when Karl plays.

19. Write the converse and contrapositive of each conditional in the previous question.

20. Let $\dot\vee$ denote the 'exclusive or' that corresponds to the English expression "either one or the other but not both". Construct a truth table for this connective.

21. Express $\phi \dot\vee \psi$ in terms of the basic combinators $\wedge, \vee, \neg$.

22. Which of the following pairs of propositions are equivalent?

   (a) $\neg(P \vee Q)$ , $\neg P \wedge \neg Q$

   (b) $\neg P \vee \neg Q$ , $\neg(P \vee \neg Q)$

   (c) $\neg(P \wedge Q)$ , $\neg P \vee \neg Q$

   (d) $\neg(P \Rightarrow (Q \wedge R))$ , $\neg(P \Rightarrow Q) \vee \neg(P \Rightarrow R)$

   (e) $P \Rightarrow (Q \Rightarrow R)$ , $(P \wedge Q) \Rightarrow R$

23. Give, if possible, an example of a true conditional sentence for which

   (a) the converse is true.

   (b) the converse is false.

   (c) the contrapositive is true.

   (d) the contrapositive is false.

24. You are in charge of a party where there are young people. Some are drinking alcohol, others soft drinks. Some are old enough to drink alcohol legally, others are under age. You are responsible for ensuring that the drinking laws are not broken, so you have asked each person to put his or her photo ID on the table. At one table are four young people. One person has a beer, another has a Coke, but their IDs happen to be face down so you cannot see their ages. You can, however, see the IDs of the other two people. One is under the drinking age, the other is above it. Unfortunately, you are not sure if they are drinking Seven-up or vodka and tonic. Which IDs and/or drinks do you need to check to make sure that no one is breaking the law?

25. Compare the logical structure of the previous question with Wason's problem (Exercise 2.3.4(8)). Comment on your answers to those two questions. In particular, identify any logical rules you used in solving each problem, say which one was easier, and why you felt it was easier.

## 2.4 Quantifiers

There are two more (mutually related) language constructions that are fundamental to expressing and proving mathematical facts, and which mathematicians therefore have to be precise about: the two *quantifiers*:

$$\textit{there exists} \quad , \quad \textit{for all}$$

The word quantifier is used in a very idiosyncratic fashion here. In normal use it means specifying the number or amount of something. In mathematics it is used to refer to the two extremes: *there is at least one* and *for all*. The reason for this restricted use is the special nature of mathematical truths. The majority of mathematical theorems—the core of mathematics when viewed as a subject in its own right (as opposed to a set of tools used in other disciplines and walks of life)—are of one of the two forms

- There is an object $x$ having property $P$

- For all objects $x$, property $P$ holds.

I'll take these one at a time. A simple example of an existence statement is:

*The equation $x^2 + 2x + 1 = 0$ has a real root.*

The existential nature of this assertion can be made more explicit by re-writing it in the form:

*There exists a real number $x$ such that $x^2 + 2x + 1 = 0$.*

Mathematicians use the symbol

$$\exists x$$

to mean

*there exists an $x$ such that ...*

Using this notation, the above example would be written symbolically as:

$$\exists x \, [x^2 + 2x + 1 = 0]$$

The symbol $\exists$ is called the *existential quantifier*. As you may have suspected, the back-to-front E comes from the word "Exists".

One obvious way to prove an existence statement is to find an object that satisfies the expressed condition. In this case, the number $x = -1$ does the trick. (It's the only number that does, but one is enough to satisfy an existence claim.)

Not all true existence claims are proved by finding a requisite object. Mathematicians have other methods for proving statements of the form $\exists x P(x)$. For example, one way to prove that the equation $x^3 + 3x + 1 = 0$ has a real root is to note that the curve $y = x^3 + 3x + 1$ is continuous (intuitively, the graph is an unbroken line), that the curve is below the $x$-axis when $x = -1$ and above the $x$-axis when $x = 1$, and hence must (by continuity) cross the $x$-axis somewhere between those two values of $x$. The value of $x$ when it crosses the $x$-axis will be a solution to the given equation. So we have proved that there is a solution without actually finding one. (It takes no small amount of fairly deep mathematics to turn this intuitively simple argument into a totally rigorous proof, but the general idea as I just explained it does work.)

*Exercise 2.4.1*

The same kind of argument I just outlined to show that the cubic equation $y = x^3 + 3x + 1$ has a real root, can be used to prove the "Wobbly Table Theorem." Suppose you are sitting in a restaurant at a perfectly square table, with four identical legs, one at each corner. Because the floor is uneven, the table wobbles. One solution is to fold a small piece of paper and insert it under one leg until the table is stable. But there is another solution. Simply by rotating the table you will be able to position it so it does not wobble. Prove this. [WARNING: This is a thinking-outside-the-box question. The solution is simple, but it can take a lot of effort before you find it. This would be an unfair question on a timed exam but is a great puzzle to keep thinking about until you hit upon the right idea.]

Sometimes it is not immediately obvious that a statement is an existence assertion. In fact, many mathematical statements that do not look like existence statements on the surface turn out to be precisely that when you work out what they mean. For example, the statement

$$\sqrt{2} \text{ is rational}$$

expresses an existence claim. You see that when you unpack its meaning and write it in the form

*There exist* natural numbers $p$ and $q$ such that $\sqrt{2} = p/q$.

Using the existential quantifier symbol, we might write this as

$$\exists p \, \exists q \, (\sqrt{2} = p/q)$$

This would be fine provided we specified in advance that the variables $p$ and $q$ refer to whole numbers. Sometimes the context in which we work guarantees that everyone knows what kinds of entities the various symbols refer to. But that is (very) often not the case. So we extend the quantifier notation by specifying the kind of entity under consideration. In this example, we would write

$$(\exists p \in \mathcal{N})(\exists q \in \mathcal{N})(\sqrt{2} = p/q)$$

This uses set-theoretic notation that you are probably familiar with. $\mathcal{N}$ denotes the set of natural numbers (i.e., positive whole numbers) and $p \in \mathcal{N}$ means "$p$ is an element (or member) of the set $\mathcal{N}$." See the appendix for a brief summary of the set theory required for this book.

Note that I did not write the above formula as $(\exists p, q \in \mathcal{N})(\sqrt{2} = p/q)$. You often see experienced mathematicians writing expressions like this, but it is definitely not recommended for beginners. Most mathematical statements involve a whole string of quantifiers, and as we'll see, it can get very tricky manipulating the expression in the course of a mathematical argument, so it is safer to stick to the "one variable per quantifier" rule. For the most part, I shall do that throughout this book.

The above statement, $(\exists p \in \mathcal{N})(\exists q \in \mathcal{N})(\sqrt{2} = p/q)$, turns out to be false. The number $\sqrt{2}$ is not rational. I'll give the proof later, but before I do, you might want to see if you can prove it yourself. The argument is only a few lines long, but it involves a clever idea. Chances are, you won't discover it, but if you do, it will definitely make your day! Give it an hour or so.

Incidentally, one feature you need to get used to in mastering college mathematics, or more generally what I am calling mathematical thinking, is the length of time you may need to spend on one particular detail. High school mathematics courses (particularly in the US) are generally put together so that most problems can be done in a few minutes, with the goal of covering an extensive curriculum. At college, there is less material to cover, but the aim is to cover it in more depth. That means you have to adjust to the slower pace, *with a lot more thinking and less doing*. At first, this comes hard, since thinking without seeming to be making progress is initially frustrating. But it's very much like learning to ride a bike. For a long time, you keep falling (or relying on training wheels), and it seems you'll never "get it." Then suddenly, one day, you find you can do it, and you cannot understand why it took so long to get there. But that long period of repeated falling was essential to your body learning how to do it. Training your mind to think mathematically about various kinds of problems is very much like that.

The remaining piece of language we need to examine and make sure we fully comprehend is the *universal quantifier*, which asserts that something holds *for all* $x$. We use the symbol

$$\forall x$$

to mean

*for all $x$ it is the case that ...*

The symbol $\forall$ is just an upside-down A, coming from the word "All".

For example, to say that the square of any real number is greater than or equal to 0, we might write

$$\forall x \, (x^2 \geq 0)$$

As before, this would be fine, provided we specified in advance that the variable $x$ refers to real numbers. It usually does, of course. But to be sure, we can modify the notation to make it crystal clear and unambiguous:

$$(\forall x \in \mathcal{R})(x^2 \geq 0)$$

We would read this as "For all real numbers $x$, the square of $x$ is greater than or equal to 0."

Most statements in mathematics involve combinations of both kinds of quantifier. For instance, the assertion that there is no largest natural number requires two quantifiers, thus:

$$(\forall m \in \mathcal{N})(\exists n \in \mathcal{N})(n > m)$$

This reads: for all natural numbers $m$ it is the case that there exists a natural number $n$ such that $n$ is greater than $m$.

Notice that the order in which quantifiers appear can be of paramount importance. For example, if we switch the order in the above we get

$$(\exists n \in \mathcal{N})(\forall m \in \mathcal{N})(n > m)$$

This asserts that there is a natural number which exceeds all natural numbers—an assertion that is clearly false!

Now it should be clear why we need to avoid using language the way the American Melanoma Foundation writer did when crafting that statement *One American dies of melanoma almost every hour.* That sentence has the logical form

$$\exists A \, \forall H [A \text{ dies in hour } H]$$

when what is meant is

$$\forall H \, \exists A [A \text{ dies in hour } H].$$

*Exercises 2.4.2*

1. Express the following as existence assertions. (Feel free to use a mix of symbols and words.)

   (a) The equation $x^3 = 27$ has a natural number solution.

   (b) 1,000,000 is not the largest natural number.

   (c) The natural number $n$ is not a prime.

2. Express the following as 'for all' assertions (using symbols and words):

   (a) The equation $x^3 = 28$ does not have a natural number solution.

   (b) 0 is less than every natural number.

   (c) the natural number $n$ is a prime.

3. Express the following in symbolic form, using quantifiers for people:

   (a) Everybody loves somebody.

   (b) Everyone is tall or short.

   (c) Everyone is tall or everyone is short.

   (d) Nobody is at home.

   (e) If John comes, all the women will leave.

   (f) If a man comes, all the women will leave.

4. Express the following using quantifiers that refer (only) to the sets $\mathcal{R}$ and $\mathcal{N}$:

   (a) The equation $x^2 + a = 0$ has a real root for any real number $a$.

   (b) The equation $x^2 + a = 0$ has a real root for any negative real number $a$.

   (c) Every real number is rational.

(d) There is an irrational number.

(e) There is no largest irrational number. (This one looks quite complicated.)

5. Let $C$ be the set of all cars, let $D(x)$ mean that $x$ is domestic, and let $M(x)$ mean that $x$ is badly made. Express the following in symbolic form using these symbols:

   (a) All domestic cars are badly made.

   (b) All foreign cars are badly made.

   (c) All badly made cars are domestic.

   (d) There is a domestic car that is not badly made.

   (e) There is a foreign car that is badly made.

6. Express the following sentence symbolically, using only quantifiers for real numbers, logical connectives, the order relation $<$, and the symbol $Q(x)$ having the meaning '$x$ is rational':

   *There is a rational number between any two unequal real numbers.*

7. Express the following famous statement (by Abraham Lincoln) using quantifiers for people and times: "You may fool all the people some of the time, you can even fool some of the people all of the time, but you cannot fool all of the people all the time."

8. A US newspaper headline read, "A driver is involved in an accident every six seconds." Let $x$ be a variable to denote a driver, $t$ a variable for a six-second interval, and $A(x, t)$ the property that $x$ is in an accident during interval $t$. Express the headline in logical notation.

In mathematics (and in everyday life), you often find yourself having to negate a statement involving quantifiers. Of course, you can do it simply by putting a negation symbol in front. But often that's not enough; you need to produce a *positive* assertion, not a *negative* one. The examples I'll give should make it clear what I mean by "positive" here, but roughly speaking, a positive statement is one that says *what is*, rather than *what is not*. In practice, a positive statement is one that contains no negation symbol, or else one in which any negation symbols are as far inside the statement as is possible without the resulting expression being unduly cumbersome.

Let $A(x)$ denote some property of $x$. (For example, $A(x)$ could say that $x$ is a real root of the equation $x^2 + 2x + 1 = 0$.) I'll show that

$$\neg[\forall x A(x)] \text{ is equivalent to } \exists x[\neg A(x)]$$

For example,

*It is not the case that all motorists run red lights*

is equivalent to

*There is a motorist who does not run red lights.*

With a familiar example like this, the equivalence is obvious. The general proof requires nothing more than formulating this general understanding in a generic, abstract form. If the following seems at all mysterious, the explanation is undoubtedly that you are simply not used to reasoning in a decontextualized, abstract manner. If you are working through this book in preparation for taking college math courses, then you will need to master abstract reasoning as soon as possible. On the other hand, if your goal is simply to improve your analytic reasoning skills for everyday use, then it is probably enough to replace the abstract symbols by specific, simple examples (as I just did) and work through them, though mastery of abstraction definitely helps everyday reasoning, by highlighting the underlying logic on which all reasoning depends.

Now for the abstract verification. We begin by assuming that $\neg[\forall x A(x)]$. That is, we assume it is not the case that $\forall x A(x)$ is true. Well, if it is not the case that all $x$ satisfy $A(x)$, what must happen is that at least one of the $x$ must fail to satisfy $A(x)$. In other words, for at least one $x$, $\neg A(x)$ must be true. In symbols, this can be written $\exists x[\neg A(x)]$. Hence $\neg[\forall x A(x)]$ implies $\exists x[\neg A(x)]$.

Now suppose $\exists x[\neg A(x)]$ Thus there will be an $x$ for which $A(x)$ fails. Hence $A(x)$ does not hold for every $x$. (It fails for the $x$ where it fails!) In other words, it is false that $A(x)$ holds for all $x$. In symbols, $\neg[\forall x A(x)]$. Thus $\exists x[\neg A(x)]$ implies $\neg[\forall x A(x)]$.

Taken together, the two implications just established produce the claimed equivalence.

## Exercises 2.4.3

1. Show that $\neg[\exists x A(x)]$ is equivalent to $\forall x[\neg A(x)]$.

2. Give an everyday example to illustrate this equivalence, and verify it by an argument specific to your example.

Now we are in a position to carry out a proper analysis of our earlier problem about domestic cars, where we want to negate the statement

*All domestic cars are badly made.*

Let us formulate this symbolically using the notation of Exercise 2.4.2(5). If you got part (a) of that question correct, you should have the formulation

$$(\forall x \in C)[D(x) \Rightarrow M(x)]$$

Negating this gives

$$(\exists x \in C)\neg[D(x) \Rightarrow M(x)]$$

(One common cause of confusion. Why do we not say $(\exists x \notin C)$? The answer is that the '$\in C$' part simply tells us which kind of $x$ we are to consider. Since our original statement concerns domestic cars, so will its negation.)

Consider now the part

$$\neg[D(x) \Rightarrow M(x)]$$

We have seen already that this is equivalent to

$$D(x) \wedge (\neg M(x))$$

Hence for our negated statement (in positive form now) we get

$$(\exists x \in C)[D(x) \wedge (\neg M(x))]$$

In words, there is a car that is domestic and is not badly made; i.e., there is a domestic car that is not badly made.

We can also obtain this result directly as follows, without going through the above symbolic manipulations.

If it is not the case that all domestic cars are badly made, then it must be the case that at least one of them fails to be badly made. Hence, as this argument reverses, the required negation is that at least one domestic car is not badly made.

The issue discussed above causes problems for enough beginners to warrant some further examples.

The first is about natural numbers. Hence all variables will refer to members of the set $\mathcal{N}$. Let $P(x)$ denotes the property "$x$ is a prime" and $O(x)$ the property "$x$ is odd". Consider the sentence

$$\forall x[P(x) \Rightarrow O(x)]$$

This says that all primes are odd, which is false. (Why? How would you prove that?) The negation of this sentence will have the (positive) form

$$\exists x[P(x) \wedge \neg O(x)]$$

To get to this form, you start with

$$\neg \forall x[P(x) \Rightarrow O(x)]$$

which is equivalent to

$$\exists x \neg [P(x) \Rightarrow O(x)]$$

and that in turn is equivalent to

$$\exists x[P(x) \nRightarrow O(x)]$$

which we can reformulate as

$$\exists x[P(x) \wedge \neg O(x)]$$

Thus the $\forall$ becomes a $\exists$ and the $\Rightarrow$ becomes a $\wedge$. In words, the negation reads "There is a prime that is not odd," or more colloquially, "There is an even prime." This is, of course, true. (Why? How would you prove this?)

Viewed as a symbolic procedure, what I did above was move the negation symbol successively inside the expression, adjusting the logical connectives appropriately as I did. As you will have suspected, it is possible to write down a list of symbol-manipulation rules for doing this kind of thing. That would be useful if you wanted to write a computer program to carry out logical reasoning. But our goal here is to develop mathematical thinking skills. The symbolic examples are merely a way of achieving that, in a manner that is particularly useful for college mathematics students. Thus, I would strongly recommend that you approach every problem in terms of what it means, using its own language.

If we modify the original sentence to read

$$(\forall x > 2)[P(x) \Rightarrow O(x)]$$

(i.e., all primes greater than 2 are odd, which is true) then the negation of this sentence can be written as

$$(\exists x > 2)[P(x) \wedge \neg O(x)]$$

(i.e., there is an even prime bigger than 2), which is false.

One thing to notice about this example is that the quantifier $(\forall x > 2)$ changes to $(\exists x > 2)$, *not* to $(\exists x \leq 2)$. Likewise, negation of the quantifier $(\exists x > 2)$ leads to $(\forall x > 2)$, *not* to $(\forall x \leq 2)$. You should make sure you understand the reason for this behavior.

## Exercise 2.4.4

Prove that the statement

*There is an even prime bigger than 2*

is false.

For another example, suppose we are talking about people, so $x$ denotes an arbitrary person. Let $P(x)$ be the property of being a player for a certain sports team and $H(x)$ the property of being healthy. Then the sentence

$$\exists x \, [P(x) \wedge \neg H(x)]$$

expresses the claim that there is an unhealthy player. Negating this gives

$$\forall x \, [\neg P(x) \vee H(x)]$$

This is a bit unnatural to read in English, but by virtue of the way we defined $\Rightarrow$, it can be rewritten as

$$\forall x \, [P(x) \Rightarrow H(x)]$$

and this has the natural reading that "all players are healthy".

Here is another mathematical example, where the variables denote members of the set $\mathcal{Q}$ of all rationals. Consider the sentence

$$\forall x \,[x > 0 \Rightarrow \exists y(xy = 1)]$$

This says that every positive rational has a multiplicative inverse (which is true). The negation of this sentence (which will be false) works out as follows.

$$\neg \forall x \,[x > 0 \Rightarrow \exists y(xy = 1)] \quad \Leftrightarrow \quad \exists x \,[x > 0 \wedge \neg \exists y(xy = 1)]$$
$$\Leftrightarrow \quad \exists x \,[x > 0 \wedge \forall y(xy \neq 1)]$$

In words, there is a positive rational $x$ with the property that no $y$ exists such that $xy = 1$, i.e., there is a positive rational with no multiplicative inverse.

The above examples illustrate a feature of quantification that is sufficiently common to warrant systematic development. Associated with any use of quantifiers there is what is known as a *domain of quantification*: the collection of all objects that the quantifiers refer to. This may be the collection of all real numbers, the collection of all natural numbers, the collection of all complex numbers, or some other collection.

In many cases, the overall context determines the domain. For instance, if we are studying real analysis, then unless otherwise mentioned it may safely be assumed that any quantifier refers to the real numbers. But on occasions there is no alternative but to be quite explicit as to what is the domain under discussion.

To illustrate how it can sometimes be important to specify the domain, consider the mathematical statement
$$\forall x \exists y \,(y^2 = x)$$

This is true for the domain $\mathcal{C}$ of complex numbers but not true when the domain is $\mathcal{R}$.

At the risk of confusing you, I should mention that, in practice, mathematicians often omit not only explicit mention of the domain of quantification (leaving it to the context to indicate what the variable denotes), but in the case of universal quantification, all mention of the quantifier at all, writing expressions such as

$$x \geq 0 \Rightarrow \sqrt{x} \geq 0$$

when what is meant is
$$(\forall x \in \mathcal{R})[x \geq 0 \Rightarrow \sqrt{x} \geq 0]$$

The former is known as *implicit quantification*. Although I do not use this convention in this book, implicit quantification is fairly common, so you should be aware of it.

Care has to be exercised when quantifiers are combined with the logical combinators $\wedge, \vee$, etc.

As an illustration of the various pitfalls that can arise, suppose the domain under discussion is the set of natural numbers. Let $E(x)$ be the statement '$x$ is even', and let $O(x)$ be the statement '$x$ is odd'.

The statement

$$\forall x[E(x) \vee O(x)]$$

says that for every natural number $x$, $x$ is either even or odd (or both). This is clearly true.

On the other hand, the statement

$$\forall x E(x) \vee \forall x O(x)$$

is false, since it asserts that either all natural numbers are even or else all natural numbers are odd (or both), whereas in fact neither of these alternatives is the case.

Thus, in general you cannot "move a $\forall x$ inside brackets." More precisely, if you do, you can end up with a very different statement, not equivalent to the original one.

Again, the statement

$$\exists x[E(x) \wedge O(x)]$$

is false, since it claims that there is a natural number that is both even and odd, whereas the statement

$$\exists x E(x) \wedge \exists x O(x)$$

claims that there is a natural number that is even and there is a natural number that is odd, which is true.

Thus "moving a $\exists x$ inside brackets" can also lead to a statement that is not equivalent to the original one.

Notice that although the last statement above uses the same variable $x$ in both parts of the conjunction, the two conjuncts operate separately.

You should make sure you fully appreciate the distinction between *all* the above example statements involving quantifiers.

Very often, in the course of an argument, we use quantifiers that are restricted to a smaller collection than the original domain. For example, in real analysis (where the unspecified domain is usually the set $\mathcal{R}$ of all real numbers) we often need to talk about "all positive numbers" or "all negative numbers", and in number theory (where the unspecified domain is the set $\mathcal{N}$ of all natural numbers) we have quantifiers such as "for all prime numbers", etc.

One way to handle this has been done already. We can modify the quantifier notation, allowing quantifiers of the form

$$(\forall x \in A) \quad , \quad (\exists x \in A)$$

where $A$ is some subcollection of the domain.

Another way is to specify the objects being quantified within the non-quantifier part of the formula. For example, suppose the domain under discussion is the set of all animals. Thus, any variable $x$ is assumed to denote an animal. Let $L(x)$ mean that "$x$ is a leopard" and let $S(x)$ mean that "$x$ has spots". Then the sentence "All leopards have spots" can be written like this:

$$\forall x[L(x) \Rightarrow S(x)]$$

In English, this reads literally as: "For all animals $x$, if $x$ is a leopard *then* $x$ has spots". This is rather cumbersome English, but the mathematical version turns out to be preferable to using a modified quantifier of the form $(\forall x \in \mathcal{L})$ where $\mathcal{L}$ denotes the set of all leopards, since a mathematical argument where quantifiers refer to different domains could easily lead to confusion and error.

Beginners often make the mistake of rendering the original sentence "All leopards have spots" as

$$\forall x[L(x) \wedge S(x)]$$

In English, what this says is: "For all animals $x$, $x$ is both a leopard and has spots". Or, smoothing out the English a bit, "All animals are leopards and have spots". This is obviously false; not all animals are leopards, for one thing.

Part of the reason for the confusion is probably the fact that the mathematics goes differently in the case of existential sentences. For example, consider the sentence "There is a horse that has spots". If we let $H(x)$ mean that "$x$ is a horse", then this sentence translates into the mathematical sentence

$$\exists x[H(x) \wedge S(x)]$$

Literally: "There is an animal that is both a horse *and* has spots".

Contrast this with the sentence

$$\exists x[H(x) \Rightarrow S(x)]$$

This says that "There is an animal such that *if* it is a horse, *then* it has spots". This does not seem to say anything much, and is certainly not at all the same as saying that there is a spotty horse.

In symbolic terms, the modified quantifier notation

$$(\forall x \in \mathcal{A})\phi(x)$$

(where the notation $\phi(x)$ indicates that $\phi$ is a statement that involves the variable $x$) may be regarded as an abbreviation for the expression

$$\forall x[A(x) \Rightarrow \phi(x)]$$

where $A(x)$ is the property of $x$ being in the collection $\mathcal{A}$.

Likewise, the notation

$$(\exists x \in \mathcal{A})\phi(x)$$

may be regarded as an abbreviation for

$$\exists x[A(x) \wedge \phi(x)]$$

In order to negate statements with more than one quantifier, you could start at the outside and work inwards, handling each quantifier in turn. The overall effect is that the negation symbol moves inwards, changing each $\forall$ to an $\exists$ and each $\exists$ to a $\forall$ as it passes. Thus, for example

$$\neg[\forall x \exists y \forall z A(x, y, z)] \quad \Leftrightarrow \quad \exists x \neg[\exists y \forall z A(x, y, z)]$$
$$\Leftrightarrow \quad \exists x \forall y \neg[\forall z A(x, y, z)]$$
$$\Leftrightarrow \quad \exists x \forall y \exists z \neg[A(x, y, z)]$$

As I said before, however, the purpose of this book is to develop thinking skills, not to learn another collection of cookie-cutter rules you can apply to avoid thinking! Industrial strength mathematics problems often involve fairly complex statements. Mathematicians do sometimes use symbolic manipulations like the above to *check* their reasoning after the fact, but they invariably do the initial reasoning in terms of what the problem means, not by first translating it to a symbolic form and then cranking a symbolic manipulation procedure. The primary goal of college-level pure mathematics, remember, is *understanding*. Doing and solving (generally the only goals emphasized at high school) are secondary goals. Applying a set of procedures does not lead to understanding. Thinking about, working with, and eventually (you hope) solving the problem *in terms of what it means* does.

One further quantifier that is often useful is

*there exists a unique x such that* ...

The usual notation for this quantifier is

$$\exists!$$

This quantifier can be defined in terms of the other quantifiers, by taking

$$\exists! x \phi(x)$$

to be an abbreviation for

$$\exists x [\phi(x) \wedge \forall y [\phi(y) \Rightarrow x = y]]$$

(Make sure you understand why this last formula expresses unique existence.)

## Exercises 2.4.5

1. Translate the following sentences into symbolic form using quantifiers. In each case the assumed domain is given in parentheses.

   (a) All students like pizza. (All people)

   (b) One of my friends does not have a car. (All people)

   (c) Some elephants do not like muffins. (All animals)

  (d) Every triangle is isosceles. (All geometric figures)

  (e) Some of the students in the class are not here today. (All people)

  (f) Everyone loves somebody. (All people)

  (g) Nobody loves everybody. (All people)

  (h) If a man comes, all the women will leave. (All people)

  (i) All people are tall or short. (All people)

  (j) All people are tall or all people are short. (All people)

  (k) Not all precious stones are beautiful. (All stones)

  (l) Nobody loves me. (All people)

  (m) At least one American snake is poisonous. (All snakes)

  (n) At least one American snake is poisonous. (All animals)

2. Which of the following are true? The domain for each is given in parentheses.

  (a) $\forall x(x + 1 \geq x)$ (Real numbers)

  (b) $\exists x(2x + 3 = 5x + 1)$ (Natural numbers)

  (c) $\exists x(x^2 + 1 = 2^x)$ (Real numbers)

  (d) $\exists x(x^2 = 2)$ (Rational numbers)

  (e) $\exists x(x^2 = 2)$ (Real numbers)

  (f) $\forall x(x^3 + 17x^2 + 6x + 100 \geq 0)$ (Real numbers)

  (g) $\exists x(x^3 + x^2 + x + 1 \geq 0)$ (Real numbers)

  (h) $\forall x \exists y(x + y = 0)$ (Real numbers)

  (i) $\exists x \forall y(x + y = 0)$ (Real numbers)

  (j) $\forall x \exists! y(y = x^2)$ (Real numbers)

  (k) $\forall x \exists! y(y = x^2)$ (Natural numbers)

  (l) $\forall x \exists y \forall z(xy = xz)$ (Real numbers)

  (m) $\forall x \exists y \forall z(xy = xz)$ (Prime numbers)

  (n) $\forall x \exists y(x \geq 0 \Rightarrow y^2 = x)$ (Real numbers)

  (o) $\forall x[x < 0 \Rightarrow \exists y(y^2 = x)]$ (Real numbers)

  (p) $\forall x[x < 0 \Rightarrow \exists y(y^2 = x)]$ (Positive real numbers)

3. Negate each of the symbolic statements you wrote in Question 1, putting your answers in positive form. Express each negation in natural, idiomatic English.

4. Negate each of the statements in Question 2, putting your answers in positive form.

5. Negate the following statements and put each answer into positive form:

   (a) $(\forall x \in \mathcal{N})(\exists y \in \mathcal{N})(x + y = 1)$

   (b) $(\forall x > 0)(\exists y < 0)(x + y = 0)$   (where $x, y$ are real number variables)

   (c) $\exists x (\forall \epsilon > 0)(-\epsilon < x < \epsilon)$   (where $x, \epsilon$ are real number variables)

   (d) $(\forall x \in \mathcal{N})(\forall y \in \mathcal{N})(\exists z \in \mathcal{N})(x + y = z^2)$

6. Give a negation (in positive form) of the quotation which you met in Exercise 2.4.2(7): "You may fool all the people some of the time, you can even fool some of the people all of the time, but you cannot fool all of the people all the time."

7. The standard definition of a real function $f$ being *continuous at a point* $x = a$ is

$$(\forall \epsilon > 0)(\exists \delta > 0)(\forall x)[|x - a| < \delta \Rightarrow |f(x) - f(a)| < \epsilon]$$

Write down a formal definition for $f$ being *discontinuous at $a$*. Your definition should be in positive form.

# 3

# Proofs

In the natural sciences, truth is established by empirical means, involving observation, measurement, and (the gold standard) experiment. In mathematics, truth is determined by constructing a *proof*—a logically sound argument that establishes the truth of the statement.

The use of the word "argument" here is, of course, not the more common everyday use to mean a disagreement between two people, but there is a connection in that a good proof will preemptively counter (implicitly or explicitly) all the objections (counterarguments) a reader might put forward. When professional mathematicians read a proof, they generally do so in a manner reminiscent of a lawyer cross-examining a witness, constantly probing and looking for flaws.

Learning how to prove things forms a major part of college mathematics. It is not something that can be mastered in a few weeks; it takes years. What can be achieved in a short period, and what I am going to try to help you do here, is gain some understanding of what it means to prove a mathematical statement, and why mathematicians make such a big deal about proofs.

## 3.1   What is a proof?

Proofs are constructed for two main purposes: to establish truth and to communicate to others.

Constructing or reading a proof is how we convince ourselves that some statement is true. I might have an intuition that some mathematical statement is true, but until I have proved it—or read a proof that convinces me—I cannot be sure.

But I may also have need to convince someone else. For both purposes, a proof of a statement must *explain why* that statement is true. In the first case, convincing myself, it is generally enough that my argument is logically sound and I can follow it later. In the second case, where I have to convince someone else, more is required: the proof must also provide that explanation in a manner the recipient can understand. Proofs written to convince others have to succeed communicatively as well as be logically sound. (For complicated proofs, the requirement that a mathematician can follow his or her proof a few days, weeks, months, or even years later can also be significant, so even proofs written purely for personal use need to succeed communicatively.)

The requirement that proofs must communicate explanations to intended readers can set a high bar. Some proofs are so deep and complex that only a few experts in the field can understand them. For example, for many centuries, most mathematicians believed—or at least held a strong suspicion—that for exponents $n \geq 3$, the equation $x^n + y^n = z^n$ has no whole number solutions for $x$, $y$, $z$. This was conjectured by the great French mathematician Pierre de Fermat in the seventeenth century, but it was not finally proved until 1994 when the British mathematician Andrew Wiles constructed a long and extremely deep proof. Though most mathematicians (myself included) lack the detailed domain knowledge to follow Wiles' proof themselves, it did convince the experts in the field (analytic number theory), and as a result, Fermat's ancient conjecture is now regarded as a theorem. (It was popularly known as Fermat's Last Theorem, since it was the last of several mathematical statements Fermat announced that remained to be proved.)

Fermat's Last Theorem is an unusual example, however. Most proofs in mathematics can be read and understood by any professional mathematician, though it can take days, weeks, or even months to understand some proofs sufficiently to be convinced by them. (The examples in this book are chosen to be understood by a typical reader in a few minutes, or possibly an hour or so. Examples given to college mathematics majors can usually be understood with at most a few hours' effort.)

Proving a mathematical statement is much more than gathering evidence in its favor. To give one famous example, in the mid-eighteenth century, the great Swiss mathematician Leonard Euler stated his belief that every even number beyond 2 can be expressed as a sum of two primes. This property of even numbers had been suggested to him by Christian Goldbach, and became known as the Goldbach Conjecture. It is possible to run computer programs to check the statement for many specific even numbers, and to date (July 2012) it has been verified for all numbers up to and beyond $1.6 \times 10^{18}$ (1.6 quintillion). Most mathematicians believe it to be true. But it has not yet been proved.

All it would take to *disprove* the conjecture would be to find a single even number $n$ for which it could be shown that no two primes sum to $n$.

Incidentally, mathematicians do not regard the Goldbach Conjecture as important. It has no known applications or even any significant consequences within mathematics. It has become famous solely because it is easy to understand, was endorsed by Euler, and has resisted all attempts at solution for over 250 years.

Whatever you may have been told at school, there is no particular format that an argument has to have in order to count as a proof. The one absolute requirement is that it is a logically sound piece of reasoning that establishes the truth of some statement. An important secondary requirement is that it is expressed sufficiently well that an intended reader can, perhaps with some effort, follow the reasoning. In the case of professional mathematicians, the intended reader is usually another professional with expertise in the same area of mathematics; proofs written for students or laypersons generally have to supply more explanations.

This means that in order to construct a proof, you have to be able to determine what constitutes a logically sound argument that convinces not just yourself, but also an intended reader. Doing that is not something you can reduce to a list of rules. Constructing mathematical proofs is one of the most creative acts of the human mind, and relatively few are capable of truly original proofs. But with some effort, any reasonably intelligent person can master the basics. That's my goal here.

Euclid's proof that there are infinitely many primes, which I gave in Chapter 2, is a good example of a proof that requires an unusual insight. There were two creative ideas in that argument. One was to adopt the strategy of showing that an enumeration of the primes up to any point, $p_1, p_2, p_3, \ldots, p_n$, can always be continued (which proves infinitude in a roundabout way). The other idea was to look at that number $(p_1 \cdot p_2 \cdot p_3 \cdot \ldots \cdot p_n) + 1$. I suspect that most of us would eventually come up with the first idea; I'd like to think I would have. (As a teenager, I simply read it in a book. I wished the author had hidden the proof and challenged readers to find it for themselves, so I could have given it a shot.) But the second creative idea is a stroke of pure genius. I'd like to think I would have eventually come up with that idea too, but I am not sure I would have. This is precisely why I find Euclid's proof so pleasing, and revel in its brilliant core idea.

## 3.2 Proof by contradiction

Here is another great example of a clever proof, which illustrates a powerful strategy called "proof by contradiction." I'll lay it out in the traditional mathematical fashion of the statement of the result, labeled a "theorem," followed by its "proof." [1] But this is just an issue of style. What makes the result a *theorem* and the argument to verify it a *proof* is that the argument is logically sound and does establish the result claimed. After I have presented the argument, I'll take a look at what makes it work as a proof.

**Theorem.** The number $\sqrt{2}$ is irrational.

*Proof:* Assume, on the contrary, that $\sqrt{2}$ were rational. Then we could find natural numbers $p$ and $q$ such that

$$\sqrt{2} = p/q$$

where $p$ and $q$ have no common factors. Squaring gives

$$2 = p^2/q^2$$

which rearranges to give

$$p^2 = 2q^2$$

Thus $p^2$ is even. Hence $p$ must be even, since $\text{odd}^2 = \text{odd}$. Thus, there is a natural number $r$ such that $p = 2r$. Substituting for $p$ in the last equation gives

$$(2r)^2 = 2q^2$$

---

[1] Theorems are a Greek invention, which is why they have a Greek name. The Romans' interest in mathematics was far more practical, so there is no Latin word "theorum" in the mathematical lexicon.

i.e.,

$$4r^2 = 2q^2$$

and dividing both sides by 2 gives

$$2r^2 = q^2.$$

Thus $q^2$ is even. Hence $q$ must be even. But $p$ is even and $p$ and $q$ have no common factors, so we have a contradiction. Hence our original assumption that $\sqrt{2}$ was rational must be false. In other words, $\sqrt{2}$ must be irrational, which is what we set out to prove.    □

(Marking the end of a proof with a box, or some other symbol, is a convention to facilitate rapid reading of mathematical texts, allowing the reader easily to skip over the proofs on first reading.)

Many instructors give this theorem as an introductory illustration of mathematical proof. They do so because it is great on several levels.

First, the result itself has enormous historical significance. When the ancient Greeks made this discovery, showing that there were geometric lengths that could not be measured by their numbers, it caused a crisis in their mathematics, and it was not until two thousand years later, toward the latter part of the nineteenth century, that mathematicians finally developed a concept of number (the real number system) adequate for measuring all geometric lengths.

Second the proof is very short. Third, it uses only elementary ideas about positive whole numbers. Fourth, it uses a very common approach. Finally, it uses a very clever idea.

Let's start with the approach. It is an example of a general method called "proof by contradiction." You want to prove some statement $\phi$. To that end, you begin by assuming $\neg\phi$. You then reason until you establish something that is obviously false. Often, this takes the form of deducing both a statement $\psi$ and its negation $\neg\psi$. Provided the reasoning is correct, there is no way that you could deduce a false consequence starting from a true assumption. Hence, your original assumption of $\neg\phi$ must be false. In other words, $\phi$ must be true.

Another way to look at this is as a special case of proof by way of the contrapositive. As we saw in Exercise 2.3.5(12), $\neg\phi \Rightarrow \theta$ is equivalent to $\neg\theta \Rightarrow \phi$. To prove $\phi$ by contradiction, you start with $\neg\phi$ and you deduce F (some false statement). That is, you establish $\neg\phi \Rightarrow F$. But this is the contrapositive of $T \Rightarrow \phi$. Hence you have proved $T \Rightarrow \phi$. Thus, by modus ponens (Exercise 2.3.4(4)), $\phi$ must be true.

Once you have become comfortable with the idea of proof by contradiction, and can see why deducing a contradiction from $\neg\phi$ does indeed constitute a proof of $\phi$, then it is impossible not to be convinced by the above argument. All you need to do is work through it line by line and ask yourself, "Is there anything in this one line that is not valid?" If you reach the last line of the proof without encountering a flaw in the reasoning, then you can be sure that $\phi$ is true.

For the proof that $\sqrt{2}$ is irrational, the entire argument hinges on the issue of even versus odd numbers. The assumption about the two numbers $p, q$ having no common factors is not a problem, since any fraction can always be written in simplest form, where the numerator and denominator do not have a common factor (other than 1).

That was a fairly lengthy discussion of such a short argument. But I know from many years of experience that beginners find this proof hard to really understand. You may think you understand it, but do you really? Let's see if you can produce a similar one. And if you can do that, let's see if you can produce a generalization? You should definitely try to do this exercise. But be prepared to spend some time at it. Remember, this is not a book about solving problems. The goal is to learn to think mathematically. And just like learning to ride a bike, to ski, or to drive a car, the only way to do that is to keep trying for yourself. Looking up the answer or being shown it by someone else does not help. It really doesn't. Look it up now and you will pay heavily later. The value comes from spending time trying to solve it for yourself.

*Exercises 3.2.1*

1. Prove that $\sqrt{3}$ is irrational.

2. Is it true that $\sqrt{N}$ is irrational for every natural number $N$?

3. If not, then for what $N$ is $\sqrt{N}$ irrational? Formulate and prove a result of the form "$\sqrt{N}$ irrational if and only if $N$ ..."

Proofs by contradiction are a common approach because they have a clear starting point. To obtain a *direct* proof of some statement $\phi$, you have to generate an argument that culminates in $\phi$. But where do you start? The only way to proceed is to try to argue successively backwards to see what chain of steps ends with $\phi$. There are many possible starting points, but just one goal, and you have to end up at that goal. That can be difficult. But with proof by contradiction, there is a clear starting point, and the proof is complete once you have deduced a contradiction—*any* contradiction. With such a wide target area, that is often a much easier task.

The proof by contradiction approach is particularly suited to establishing that a certain object does not exist; for example, that a particular kind of equation does not have a solution. You begin by assuming that such an object does exist, then you use that (assumed) object to deduce a false consequence or a pair of contradictory statements. The irrationality of $\sqrt{2}$ is a good example, since that states the non-existence of two natural numbers $p, q$ whose ratio is equal to $\sqrt{2}$.

# 3.3   Proving conditionals

Even though there is no cookie-cutter, template approach to constructing proofs, there are some guidelines, and we have just encountered two. Proof by contradiction is often a good approach when there is no obvious place to start, and in particular that makes it a useful method to prove non-existence statements. Of course, you still have to construct a proof. You've simply replaced a narrow goalpost with an unclear starting point by a much wider one with a known starting point! But like Robert Frost's fork in the trail, that choice can make all the difference.

There are a number of other guidelines. I'll tell you some, but do bear in mind that these are not templates. As long as you continue to look for templates to construct proofs, you are going to encounter significant difficulties. You have to start each new problem by analyzing the statement that you want to prove. What exactly does it say? What kind of argument might establish that claim?

For example, suppose we wish to establish the truth of a conditional

$$\phi \Rightarrow \psi$$

By the definition of the conditional, this will certainly be true whenever $\phi$ is false, so we need only consider the case when $\phi$ is true. That is, we can *assume* $\phi$. Then, for the conditional to be valid, $\psi$ must also be true.

Thus, using our assumption that $\phi$ is true, we must present an argument that demonstrates the truth of $\psi$. This, of course, accords with our everyday understanding of implication. Thus, when it comes to *proving* conditionals, the problems concerning the distinction between the conditional and real implication that we discussed do not arise.

To take a specific example, suppose we want to prove that for any given pair $x, y$ of real numbers:

$$(x \text{ and } y \text{ are rational numbers}) \Rightarrow (x + y \text{ is a rational number})$$

We start by *assuming* $x$ and $y$ are rational numbers. Then we can find integers $p, q, m, n$ such that $x = p/m, y = q/n$. Then

$$x + y = \frac{p}{m} + \frac{q}{n} = \frac{pn + qm}{mn}$$

Hence, as $pn + qm$ and $mn$ are integers, we conclude that $x + y$ is rational. The statement is proved.

## Exercise 3.3.1

Let $r$, $s$ be irrationals. For each of the following, say whether the given number is necessarily irrational, and prove your answer. (The last one is particularly nice. I'll give the solution in a moment, but you should definitely try it first.)

    1. $r + 3$      2. $5r$      3. $r + s$

    4. $rs$      5. $\sqrt{r}$      6. $r^s$

Conditionals involving quantifiers are sometimes best handled by *proving the contrapositive*, using the equivalence of $\phi \Rightarrow \psi$ with the contrapositive $(\neg\psi) \Rightarrow (\neg\phi)$.

For example, suppose that for some given unknown angle $\theta$ we wish to prove the conditional

$$(\sin\theta \neq 0) \Rightarrow (\forall n \in \mathcal{N})(\theta \neq n\pi)$$

This statement is equivalent to

$$\neg(\forall n \in \mathcal{N})(\theta \neq n\pi) \Rightarrow \neg(\sin\theta \neq 0)$$

which reduces to the positive form

$$(\exists n \in \mathcal{N})(\theta = n\pi) \Rightarrow (\sin\theta = 0)$$

This is an implication we know to be correct. This proves the original implication by virtue of what equivalence means. (To prove a statement, it is enough to prove any equivalent statement.)

To prove a biconditional (an equivalence) $\phi \Leftrightarrow \psi$, you generally prove the two conditionals $\phi \Rightarrow \psi$ and $\psi \Rightarrow \phi$. (Why is this enough?)

Occasionally, however, you might find it more natural to prove the two conditionals $\phi \Rightarrow \psi$ and $(\neg\phi) \Rightarrow (\neg\psi)$. (Why will this work?)

## Exercises 3.3.2

1. Explain why proving $\phi \Rightarrow \psi$ and $\psi \Rightarrow \phi$ estabishes the truth of $\phi \Leftrightarrow \psi$.

2. Explain why proving $\phi \Rightarrow \psi$ and $(\neg\phi) \Rightarrow (\neg\psi)$ estabishes the truth of $\phi \Leftrightarrow \psi$.

3. Prove that if five investors split a payout of $2 million, at least one investor receives at least $400,000.

4. Write down the converses of the following conditional statements:

    (a) If the Dollar falls, the Yuan will rise.

    (b) If $x < y$ then $-y < -x$. (For $x, y$ real numbers.)

    (c) If two triangles are congruent they have the same area.

    (d) The quadratic equation $ax^2 + bx + c = 0$ has a solution whenever $b^2 \geq 4a$. (Where $a, b, c, x$ denote real numbers and $a \neq 0$.)

    (e) Let $ABCD$ be a quadrilateral. If the opposite sides of $ABCD$ are pairwise equal, then the opposite angles are pairwise equal.

    (f) Let $ABCD$ be a quadrilateral. If all four sides of $ABCD$ are equal, then all four angles are equal.

    (g) If $n$ is not divisible by 3 then $n^2 + 5$ is divisible by 3. (For $n$ a natural number.)

5. Discounting the first example, which of the statements in the previous exercise are true, for which is the converse true, and which are equivalent? Prove your answers.

6. Let $m$ and $n$ be integers. Prove that:

    (a) If $m$ and $n$ are even, then $m + n$ is even.

    (b) If $m$ and $n$ are even, then $mn$ is divisible by 4.

    (c) If $m$ and $n$ are odd, then $m + n$ is even.

    (d) If one of $m, n$ is even and the other is odd, then $m + n$ is odd.

    (e) If one of $m, n$ is even and the other is odd, then $mn$ is even.

7. Prove or disprove the statement "An integer $n$ is divisible by 12 if and only if $n^3$ is divisible by 12."

8. If you have not yet solved Exercise 3.3.1(6), have another attempt, using the hint to try $s = \sqrt{2}$.

## 3.4   Proving quantified statements

The most obvious way to prove an existence statement $\exists x A(x)$ is to find a particular object $a$ for which $A(a)$. For example, to prove that an irrational number exists it is enough to show that $\sqrt{2}$ is irrational. But sometimes you have to adopt a more indirect route. For example, such is the case with the last part of Exercises 3.3.1, which I promised I'd come back to. Here it is. (If you did not get it yet, you might want to give it one more try before you read on.)

**Theorem.** There are irrationals $r$, $s$ such that $r^s$ is rational.

*Proof:* We consider two cases.

Case 1. If $\sqrt{2}^{\sqrt{2}}$ is rational, we can take $r = s = \sqrt{2}$ and the theorem is proved.

Case 2. If $\sqrt{2}^{\sqrt{2}}$ is irrational, we can take $r = \sqrt{2}^{\sqrt{2}}$, $s = \sqrt{2}$, and then

$$(\sqrt{2}^{\sqrt{2}})^{\sqrt{2}} = (\sqrt{2})^{(\sqrt{2}\cdot\sqrt{2})} = (\sqrt{2})^2 = 2$$

and again the theorem is proved.   □

Note that in the above proof, we do not know which of the two possibilities holds. We did not produce two specific irrationals $r$, $s$ such that $r^s$ is rational. We simply showed that such a pair exists. Our proof is an example of *proof by cases*, which is another technique that can be useful.

Next, let's take a look at how to prove a universal statement $\forall x A(x)$. One possibility is to take an arbitrary $x$ and show that it must satisfy $A(x)$. For instance, suppose we wish to prove the assertion

$$(\forall n \in \mathcal{N})(\exists m \in \mathcal{N})(m > n^2)$$

We can do this as follows.

Let $n$ be an arbitrary natural number. Then $n^2$ is a natural number. Hence $m = n^2 + 1$ is a natural number. Since $m > n^2$, this shows that

$$(\exists m \in \mathcal{N})(m > n^2)$$

This is a proof because our original $n$ was quite *arbitrary*. We said nothing at all about $n$: it could be any natural number. Hence the argument is valid for *all* $n$ in $\mathcal{N}$. This is not the same as picking a *particular* $n$. If we had randomly chosen, say, $n = 37$, the proof would not have been valid—even though we had chosen this $n$ quite at random. For instance, suppose we wanted to prove

$$(\forall n \in \mathcal{N})(n^2 = 81)$$

By picking at random a particular $n$ we might happen to pick $n = 9$. But this does not prove the statement of course, because our choice was an arbitrary choice (albeit an unlucky one as far as our goal is concerned) of a *particular* $n$, and not a choice of an *arbitrary* $n$.

In practice what this amounts to is that whenever we start a proof by saying, "Let $n$ be arbitrary", we use the symbol $n$ throughout the proof, and assume also that the value of $n$ remains constant throughout, but we make absolutely no restriction on what the value of $n$ is.

Statements of the form $\forall x A(x)$ are sometimes proved by the method of contradiction. By assuming $\neg \forall x A(x)$ we obtain an $x$ such that $\neg A(x)$ (because $\neg \forall x A(x)$ is equivalent to $\exists x \neg A(x)$). Now we have a place to start. The difficulty is finding the finish (i.e., the contradiction).

## Exercises 3.4.1

1. Prove or disprove the statement "All birds can fly."

2. Prove or disprove the claim $(\forall x, y \in \mathcal{R})[(x - y)^2 > 0]$.

3. Prove that between any two unequal rationals there is a third rational.

4. Say whether each of the following is true or false, and support your decision by a *proof*:

   (a) There exist real numbers $x$ and $y$ such that $x + y = y$.
   (b) $\forall x \exists y (x + y = 0)$ (where $x, y$ are real number variables).
   (c) $(\exists m \in \mathcal{N})(\exists n \in \mathcal{N})(3m + 5n = 12)$.
   (d) For all integers $a, b, c$, if $a$ divides $bc$ (without remainder), then either $a$ divides $b$ or $a$ divides $c$.
   (e) The sum of any five consecutive integers is divisible by 5 (without remainder).
   (f) For any integer $n$, the number $n^2 + n + 1$ is odd.

(g) Between any two distinct rational numbers there is a third rational number.

(h) For any real numbers $x, y$, if $x$ is rational and $y$ is irrational, then $x + y$ is irrational.

(i) For any real numbers $x, y$, if $x + y$ is irrational, then at least one of $x, y$ is irrational.

(j) For any real numbers $x, y$, if $x + y$ is rational, then at least one of $x, y$ is rational.

5. Prove or disprove the claim that there are integers $m, n$ such that $m^2 + mn + n^2$ is a perfect square.

6. Prove that for any positive $m$ there is a positive integer $n$ such that $mn + 1$ is a perfect square.

7. Show that there is a quadratic $f(n) = n^2 + bn + c$, with positive integer coefficients $b$, $c$, such that $f(n)$ is composite (i.., not prime) for all positive integers $n$.

8. Prove that for any finite collection of points in the plane, not all collinear, there is a triangle having three of the points as its vertices, which contains none of the other points in its interior.

9. Prove that if every even natural number greater than 2 is a sum of two primes (the Goldbach Conjecture), then every odd natural number greater than 5 is a sum of three primes.

There are other possibilities for proving universally quantified statements. In particular, statements of the form

$$(\forall n \in \mathcal{N}) A(n)$$

where the quantification is over all natural numbers, are often proved by a method known as *induction*.

## 3.5   Induction proofs

Number theory is one of the most important branches of mathematics. It studies the properties of the natural numbers, 1, 2, 3, .... We'll look at some elementary topics from number theory in the next chapter, but for now it provides good examples of induction proofs. For example, suppose we wanted to prove that for any natural number $n$:

$$1 + 2 + \ldots + n = \tfrac{1}{2}n(n + 1)$$

As a first step, we might check the first few cases to see if the result holds for them:

$n = 1$:  $1 = \frac{1}{2}(1)(1+1)$. Both sides equal 1. Correct.

$n = 2$:  $1+2 = \frac{1}{2}(2)(2+1)$. Both sides equal 3. Correct.

$n = 3$:  $1+2+3 = \frac{1}{2}(3)(3+1)$. Both sides equal 6. Correct.

$n = 4$:  $1+2+3+4 = \frac{1}{2}(4)(4+1)$. Both sides equal 10. Correct.

$n = 5$:  $1+2+3+4+5 = \frac{1}{2}(5)(5+1)$. Both sides equal 15. Correct.

After doing one or two more cases and finding they all check out, you might suspect that the formula is indeed valid for all $n$. But a long run of confirmatory instances on its own does not constitute a proof.

For example, try working out values of the polynomial $P(n) = n^2 - n + 41$, for $n = 1, 2, 3, \ldots$. You will find that every value you calculate is a prime number. Unless, that is, you go as far as $n = 41$. $P(n)$ is indeed prime for all $n$ from 1 through 40, but $P(41) = 1681 = 41^2$. This particular prime-generating polynomial was discovered by Euler in 1772.

On the other hand, the series of computations we did to check the formula for the sums of the natural numbers provides more than just numbers that agree. As we work out one case after another, we start to sense a pattern. The method of mathematical induction is a valid method of proof that works by identifying a repeating pattern. Intuitively, what we need to show is that, however far we have established the desired result, we can always prove it for one more step. Let's make that precise.

METHOD OF PROOF BY MATHEMATICAL INDUCTION. To prove a statement of the form

$$(\forall n \in \mathcal{N})A(n)$$

by induction, you verify the following two statements:

(1)  $A(1)$    (Initial step)

(2)  $(\forall n \in \mathcal{N})[A(n) \Rightarrow A(n+1)]$    (Induction step)

That this implies $(\forall n \in \mathcal{N})A(n)$ may be reasoned as follows. By (1), $A(1)$. By (2) (as a special case) we have $A(1) \Rightarrow A(2)$. Hence $A(2)$. Again, a special case of (2) is $A(2) \Rightarrow A(3)$. Hence $A(3)$. And so on right through the natural numbers.

The thing to notice about this is that neither of the two statements we actually prove is the statement we are trying to establish. What we prove are the initial case (1) and the conditional (2). The step from these two to the conclusion $(\forall n \in \mathcal{N})A(n)$ (which I just explained) is known as the *principle of mathematical induction*.

As an example, let us use the method of induction to prove that result about the sums of the natural numbers.

**Theorem.** For any $n$,

$$1 + 2 + \ldots + n = \tfrac{1}{2}n(n+1)$$

*Proof:* We first verify the result for $n = 1$. In this case, the identity reduces to $1 = \frac{1}{2}(1)(1 + 1)$, which is correct, since both sides equal 1.

We now assume the identity holds for an arbitrary $n$:

$$1 + 2 + \ldots + n = \tfrac{1}{2}n(n + 1)$$

Add $(n + 1)$ to both sides of this (assumed) identity:

$$\begin{aligned}
1 + 2 + \ldots + n + (n + 1) &= \tfrac{1}{2}n(n + 1) + (n + 1) \\
&= \tfrac{1}{2}[n(n + 1) + 2(n + 1)] \\
&= \tfrac{1}{2}[n^2 + n + 2n + 2] \\
&= \tfrac{1}{2}[n^2 + 3n + 2] \\
&= \tfrac{1}{2}[(n + 1)(n + 2)] \\
&= \tfrac{1}{2}(n + 1)((n + 1) + 1)
\end{aligned}$$

which is the formula for $n + 1$ in place of $n$.

Hence, by the principle of mathematical induction, we can conclude that the identity does indeed hold for all $n$.    $\square$

## Exercises 3.5.1

In the above proof:

1. Write down the statement $A(n)$ which is being proved by induction.

2. Write down $A(1)$, the initial step.

3. Write down the statement $(\forall n \in \mathcal{N})[A(n) \Rightarrow A(n + 1)]$, the induction step.

Although proof by induction seems intuitively obvious—that you are showing that a step-by-step process through all the natural numbers will never break down—the principle itself is actually fairly deep. (The depth comes because the conclusion is about the infinite set of all natural numbers, and issues involving infinity are rarely simple.)

Here is another example. In this case, I'll make explicit the connection to the principle of induction as I stated it, though in practice that is not necessary

**Theorem.** If $x > 0$ then for any $n \in \mathcal{N}$,

$$(1 + x)^{n+1} > 1 + (n + 1)x$$

*Proof:* Let $A(n)$ be the statement

$$(1 + x)^{n+1} > 1 + (n + 1)x$$

Then $A(1)$ is the statement

$$(1 + x)^2 > 1 + 2x$$

which is true by virtue of the binomial expansion

$$(1+x)^2 = 1 + 2x + x^2$$

together with the fact that $x > 0$.

The next step is to prove the statement

$$(\forall n \in \mathcal{N})[A(n) \Rightarrow A(n+1)]$$

To do this we take an arbitrary (!) $n$ in $\mathcal{N}$ and prove the conditional

$$A(n) \Rightarrow A(n+1)$$

To do this we assume $A(n)$ and try to deduce $A(n+1)$. We have

$$
\begin{aligned}
(1+x)^{n+2} &= (1+x)^{n+1}(1+x) \\
&> (1+(n+1)x)(1+x) \quad [\text{by } A(n)] \\
&= 1 + (n+1)x + x + (n+1)x^2 \\
&= 1 + (n+2)x + (n+1)x^2 \\
&> 1 + (n+2)x \quad [\text{since } x > 0]
\end{aligned}
$$

This proves $A(n+1)$.

Hence by induction (that is to say, by the principle of mathematical induction) the theorem is proved. $\square$

In summary, for the method of proof by induction, you wish to prove that some statement $A(n)$ is valid for all natural numbers $n$. First, you establish $A(1)$. This is usually just a matter of trivial observation. You then present an algebraic argument which establishes the conditional

$$A(n) \Rightarrow A(n+1)$$

for any $n$. In general, you do this as follows. Assume $A(n)$. Look at the statement of $A(n+1)$, and somehow try to reduce it to $A(n)$, which has been assumed true, and thereby deduce the truth of $A(n+1)$. This accomplished, the induction proof is complete, by virtue of the principle of mathematical induction.

In setting out an induction proof formally, three points should be remembered:

(1) State clearly that the method of induction is being used.

(2) Prove the case $n = 1$ (or at the very least make the explicit observation that this case is obviously true, if this is the case).

(3) (The hard part.) Prove the conditional

$$A(n) \Rightarrow A(n+1)$$

One variant of induction that sometimes arises concerns the proof of statements such as

$$(\forall n \geq n_0)A(n)$$

where $n_0$ is some given natural number. In such a case, the first step of the induction proof consists not of a verification of $A(1)$ (which may not be true) but of $A(n_0)$ (the first case). The second step of the proof consists of the proof of the statement

$$(\forall n \geq n_0)[A(n) \Rightarrow A(n+1)]$$

This is what happens in the following theorem, part of the *Fundamental Theorem of Arithmetic*.

**Theorem.** Every natural number greater than 1 is either a prime or a product of primes.

*Proof:* At first you might think that the statement to be proved by induction is

$$(\forall n \in \mathcal{N})A(n)$$

where

$$A(n) : n \text{ is either a prime or a product of primes}$$

However, as will shortly become clear, it is more convenient to replace $A(n)$ by the (stronger) statement

$B(n)$ : every natural number $m$ such that $1 < m \leq n$ is either a prime or a product of primes

So, here goes. We prove, by induction, that $B(n)$ is true for all natural numbers $n > 1$. This clearly proves the theorem.

For $n = 2$, the result is trivial: $B(2)$ holds because 2 is prime. (Notice that in this case we must start at $n = 2$ rather than the more common $n = 1$.)

Now assume $B(n)$. We deduce $B(n+1)$. Let $m$ be a natural number such that $1 < m \leq n + 1$. If $m \leq n$, then by $B(n)$, $m$ is either a prime or a product of primes. So in order to prove $B(n+1)$, we need only show that $n + 1$ itself is either a prime or a product of primes. If $n + 1$ is a prime there is nothing further to say. Otherwise, $n + 1$ is composite, which means there are natural numbers $p, q$, such that

$$1 < p, q < n + 1$$

and

$$n + 1 = pq$$

Now $p, q \leq n$ so by $B(n)$, each of $p$ and $q$ is either a prime or a product of primes. But then $n + 1 = pq$ is a product of primes. This completes the proof of $B(n+1)$.

The theorem now follows by induction. More precisely, the principle of mathematical induction yields the validity of the statement

$$(\forall n \in \mathcal{N})B(n)$$

which trivially implies the theorem. $\square$

Of course, in the above example, the conditional

$$B(n) \Rightarrow B(n+1)$$

was rather easy to establish. (Indeed, we used $B(n)$ rather than the more obvious $A(n)$ mentioned earlier, precisely in order to carry through this simple argument.) In many cases, real ingenuity is required. But do not be misled into confusing the proof by induction of the main result

$$(\forall n \in \mathcal{N})A(n)$$

with the technical subproof of the induction step

$$(\forall n \in \mathcal{N})[A(n) \Rightarrow A(n+1)]$$

Without an announcement of the fact that induction is being used, and an observation or proof that $A(1)$ is valid, no amount of technical cleverness at proving the conditional $[A(n) \Rightarrow A(n+1)]$ will amount to a proof of the statement $(\forall n \in \mathcal{N})A(n)$.

## Exercises 3.5.2

1. Use the method of induction to prove that the sum of the first $n$ odd numbers is equal to $n^2$.

2. Prove the following by induction:

   (a) $4^n - 1$ is divisible by 3.

   (b) $(n+1)! > 2^{n+3}$ for all $n \geq 5$.

3. The notation

$$\sum_{i=1}^{n} a_i$$

is a common abbreviation for the sum

$$a_1 + a_2 + a_3 + \ldots + a_n$$

For instance,

$$\sum_{r=1}^{n} r^2$$

denotes the sum

$$1^2 + 2^2 + 3^2 + \ldots + n^2$$

Prove the following by induction:

(a) $\forall n \in \mathcal{N} : \sum_{r=1}^{n} r^2 = \frac{1}{6}n(n+1)(2n+1)$

(b) $\forall n \in \mathcal{N} : \sum_{r=1}^{n} 2^r = 2^{n+1} - 2$

(c) $\forall n \in \mathcal{N} : \sum_{r=1}^{n} r.r! = (n+1)! - 1$

4. In this section, we used induction to prove the general theorem

$$1 + 2 + \ldots + n = \tfrac{1}{2}n(n+1)$$

There is an alternative proof that does not use induction, famous because Gauss used the key idea to solve a "busywork" arithmetic problem his teacher gave to the class when he was a small child at school. The teacher asked the class to calculate the sum of the first hundred natural numbers. Gauss noted that if

$$1 + 2 + \ldots + 100 = N$$

then you can reverse the order of the addition and the answer will be the same:

$$100 + 99 + \ldots + 1 = N.$$

So by adding those two equations, you get

$$101 + 101 + \ldots + 101 = 2N$$

and since there are 100 terms in the sum, this can be written as

$$100 \cdot 101 = 2N$$

and hence

$$N = \tfrac{1}{2}(100 \cdot 101) = 5,050.$$

Generalize Gauss's idea to prove the theorem without recourse to the method of induction.

# 4

# Proving results about numbers

Though the focus of this book is a particular kind of thinking (rather than any specific mathematics), the integers and the real numbers provide convenient mathematical domains (number theory and elementary real analysis, respectively) to illustrate mathematical proofs—the principal advantage from an educational perspective being that everyone has some familiarity with both number systems, yet very likely won't have been exposed to their mathematical theories.

## 4.1   The integers

Most people's experience with the whole numbers is by way of elementary arithmetic. Yet the mathematical study of the integers, looking beyond mere calculation to the abstract properties those numbers exhibit, goes back to the very beginnings of recognizable mathematics around 700 BCE. That study has grown into one of the most important branches of pure mathematics: number theory. Most college mathematics majors find that number theory is one of the most fascinating courses they take. Not only is the subject full of tantalizing problems that are easy to state but require great ingenuity to solve—if indeed they have been solved—but some of the results turn out to have applications crucial for modern life, internet security being arguably the most important. Unfortunately we'll barely scratch the surface of number theory in this book, since my goal is different. But if anything you see in this section arouses your curiosity, I would recommend that you look further. You are unlikely to be disappointed.

The mathematical interest in the integers lies not in their use in counting, but in their arithmetical system: Given any two integers you can add them, subtract one from the other, or multiply them together, and the result will always be another integer. Division is not so straightforward, and that is where things get particularly interesting. For some pairs of integers, say 5 and 15, division is possible: 15 divides by 5 to give the integer result 3. For other pairs, say 7 and 15, division is not possible unless you are prepared to allow fractional results (which takes you outside the integers).

If you restrict arithmetic to the integers, division actually leads to two numbers, a *quotient* and a *remainder*. For example, if you divide 9 by 4 you get a *quotient*

of 2 and a *remainder* of 1:
$$9 = 4 \cdot 2 + 1$$

This is a special case of our first formal theorem concerning integers: the Division Theorem. For the proof, it is convenient to revisit the idea of *absolute value*.

Given any integer $a$, let $|a|$ denote the number that results from dropping any minus sign. The formal definition specifies two cases:

$$|a| = \begin{cases} a, & \text{if } a \geq 0 \\ -a, & \text{if } a < 0 \end{cases}$$

For example, $|5| = 5$ and $|-9| = 9$.

The number $|a|$ is called the *absolute value* of $a$.

**Theorem 4.1.1 (The Division Theorem)** Let $a, b$ be integers, $b > 0$. Then there are unique integers $q, r$ such that $a = q \cdot b + r$ and $0 \leq r < b$.

*Proof:* There are two things to be proved: that $q, r$ exist with the stated properties and that such $q, r$ are unique. We prove existence first.

The idea is to look at all non-negative integers of the form $a - kb$, where $k$ is an integer, and show that one of them is less than $b$. (Any value of $k$ for which this occurs will be a suitable $q$, the value $r$ then being given by $r = a - kb$.)

Are there any such integers $a - kb \geq 0$ ? Yes, there are. Take $k = -|a|$. Then, since $b \geq 1$,

$$a - kb = a + |a| \cdot b \geq a + |a| \geq 0$$

Since such integers $a - kb \geq 0$ do exist, there will be a smallest one, of course. Call it $r$, and let $q$ be the value of $k$ for which it occurs, so that $r = a - qb$. To complete the (existence) proof, we show that $r < b$.

Suppose, on the contrary, that $r \geq b$. Then

$$a - (q + 1)b = a - qb - b = r - b \geq 0$$

Thus, $a - (q+1)b$ is a non-negative integer of the form $a - kb$. But $r$ was chosen as the smallest one, and yet $a - (q+1)b < a - qb = r$, so this situation is contradictory. Thus, it must be the case that $r < b$, as we wished to prove.

That leaves us with the proof of uniqueness of $q, r$. The idea is to show that if there are representations

$$a = qb + r = q'b + r'$$

of $a$ with $0 \leq r, r' < b$, then in fact $r = r'$ and $q = q'$.

We start by rearranging the above equation as

$$(1) \qquad r' - r = b \cdot (q - q')$$

Taking absolute values:

$$(2) \qquad |r' - r| = b \cdot |q - q'|$$

But
$$-b < -r \leq 0 \quad \text{and} \quad 0 \leq r' < b$$
which together imply that
$$-b < r' - r < b$$
or in other words
$$|r' - r| < b$$
Hence by (2),
$$b \cdot |q - q'| < b$$
This implies that
$$|q - q'| < 1$$

There is only one possibility now, namely that $q - q' = 0$, i.e., $q = q'$. It follows at once using (1) that $r = r'$. The proof is complete.   $\square$

If this is the first full-blown, rigorous proof of a theorem like this that you have encountered, you will probably need to spend some time going over it. The result itself is not deep; it is something we are all familiar with. Our focus here is on the method we use to prove, conclusively, that the Division Theorem is true for *all* pairs of integers. Time spent now making sure you understand how the above proof works—why every step is critical—will pay dividends later on when you encounter more difficult proofs.

By gaining experience with mathematical proofs of simple results like this one, which is obvious, mathematicians become confident in the method of proof and can accept results that are not at all obvious.

For example, in the late nineteenth century the famous German mathematician David Hilbert described a hypothetical hotel that has a strange property. Hilbert's Hotel, as it has become known, is the ultimate hotel in that is has infinitely many rooms. As in most hotels, the rooms are numbered using the natural numbers, 1, 2, 3, etc.

One night, all rooms are occupied, when an additional guest turns up.

"I'm sorry," says the desk clerk, "all our rooms are occupied. You will have to go somewhere else."

The guest, a mathematician, thinks for a while before saying, "There is a way you can give me a room, without having to eject any of your existing guests."

[Before I proceed with this story, you might like to see if you can see the solution the mathematician guest has seen.]

The clerk is skeptical, but he asks the mathematician to explain how he can free up a room without ejecting anyone already in the hotel.

"It's simple," the mathematician begins. "You move everyone into the next room. So the occupant of Room 1 moves into Room 2, the occupant of Room 2 moves into Room 3, and so on throughout the hotel. In general, the occupant of Room $n$ moves into Room $n+1$. When you have done that, Room 1 is empty. You put me in that room!"

The clerk thinks about it for a moment, and then has to agree that the method will work. It is indeed possible to accommodate an additional guest in a completely full hotel without having to eject anyone. The mathematician's reasoning is totally sound. And so the mathematician gets a room for the night.

The key to the Hilbert Hotel argument is that the hotel has infinitely many rooms. Indeed, Hilbert formulated the story to illustrate one of several surprising properties of infinity. You should think about the above argument for a while. You won't learn anything new about real-world hotels, but you will come to understand infinity a bit better.

The significance of understanding infinity is that it is the key to calculus, the bedrock of modern science and engineering. And one way to handle infinity is to specify procedures that describe how infinitely many steps can be carried out.

When you are satisfied you understand Hilbert's solution (or if you don't think there is anything deep going on), try the following variants.

*Exercises 4.1.1*

1. The Hilbert Hotel scenario is as before, but this time, two guests arrive at the already full hotel. How can they be accommodated (in separate rooms) without anyone having to be ejected?

2. This time, the desk clerk faces an even worse headache. The hotel is full, but an infinite tour group arrives, each group member wearing a badge that says "HELLO, I'M $N$", for $N = 1, 2, 3, \ldots$ Can the clerk find a way to give all the new guests a room to themselves, without having to eject any of the existing guests? How?

Examples like the Hilbert Hotel demonstrate the importance of rigorous proofs in mathematics. When used to verify "obvious" results like the Division Theorem, they may seem frivolous, but when the same method is applied to issues we are not familiar with (such as questions that involve infinity), rigorous proofs are the only thing we can rely on.

The Division Theorem as stated above applies only to the division of an integer $a$ by a *positive* integer $b$. More general is:

**Theorem 4.1.2 (Generalized Division Theorem)** Let $a, b$ be integers, $b \neq 0$. Then there are unique integers $q, r$ such that

$$a = q \cdot b + r \quad \text{and} \quad 0 \leq r < |b|$$

*Proof:* The case where $b > 0$ has been dealt with in Theorem 4.1.1, so assume now that $b < 0$. Since $|b| > 0$, applying Theorem 4.1.1 provides unique integers $q', r'$ such that

$$a = q' \cdot |b| + r' \quad \text{and} \quad 0 \leq r' < |b|$$

Set $q = -q', r = r'$. Then, since $|b| = -b$, we get

$$a = q \cdot b + r \quad \text{and} \quad 0 \leq r < |b|$$

as required. □

Theorem 4.1.2 is also sometimes referred to simply as the Division Theorem. In each case, the number $q$ is referred to as the *quotient* of $a$ by $b$ and $r$ is called the *remainder*.

Simple though it is, the Division Theorem yields many results that can be of assistance in computational work. For instance (and this is a very simple example), if you were faced with a search for numbers which are squares of primes, it might be helpful to know that the square of any odd number is one more than a multiple of 8. (For example, $3^2 = 9 = 8 + 1, 5^2 = 25 = 3 \cdot 8 + 1$.) To verify this fact, note that by the Division Theorem, any number can be expressed in one of the forms $4q, 4q + 1, 4q + 2, 4q + 3$, so any odd number has one of the forms $4q + 1, 4q + 3$. Squaring each of these gives

$$(4q + 1)^2 = 16q^2 + 8q + 1 = 8(2q^2 + 1) + 1$$
$$(4q + 3)^2 = 16q^2 + 24q + 9 = 8(2q^2 + 3q + 1) + 1$$

In both cases the result is one more than a multiple of 8.

In case division of $a$ by $b$ produces a remainder $r = 0$, we say that $a$ is *divisible* by $b$. That is to say, an integer $a$ is said to be *divisible* by a nonzero integer $b$ if and only if there is an integer $q$ such that $a = b \cdot q$. For example, 45 is divisible by 9 whereas 44 is not divisible by 9. The standard notation to denote that $a$ is divisible by $b$ is $b|a$. Note that, by definition, $b|a$ implies $b \neq 0$.

You should pay particular attention to the fact that the notation $b|a$ refers to a *relationship* between the two numbers $a$ and $b$. It is either true or false. It is not a notation for a number. Be careful not to confuse $b|a$ with $a/b$. (The latter *is* a notation for a number.)

The following exercises use the standard notation $\mathcal{Z}$ for the set of all integers. (The Z comes from the German word for numbers: *Zahlen.*)

## Exercises 4.1.2

1. Express as concisely and accurately as you can the relationship between $b|a$ and $a/b$.

2. Determine whether each of the following is true or false. Prove your answers.

   (a) $0|7$      (b) $9|0$          (c) $0|0$          (d) $1|1$

   (e) $7|44$      (f) $7|(-42)$      (g) $(-7)|49$      (h) $(-7)|(-56)$

   (i) $2708|569401$        (j) $(\forall n \in \mathcal{N})[2n|n^2]$    (k) $(\forall n \in \mathcal{Z})[2n|n^2]$

   (l) $(\forall n \in \mathcal{Z})[1|n]$      (m) $(\forall n \in \mathcal{N})[n|0]$        (n) $(\forall n \in \mathcal{Z})[n|0]$

   (o) $(\forall n \in \mathcal{N})[n|n]$      (p) $(\forall n \in \mathcal{Z})[n|n]$

The following theorem lists the basic properties of divisibility.

**Theorem 4.1.3** Let $a, b, c, d$ be integers, $a \neq 0$. Then:

   (i) $a|0$,  $a|a$ ;

  (ii) $a|1$ if and only if $a = \pm 1$ ;

 (iii) if $a|b$ and $c|d$, then $ac|bd$ (for $c \neq 0$) ;

 (iv) if $a|b$ and $b|c$, then $a|c$ (for $b \neq 0$) ;

  (v) $[a|b$ and $b|a]$ if and only if $a = \pm b$ ;

 (vi) if $a|b$ and $b \neq 0$, then $|a| \leq |b|$ ;

(vii) if $a|b$ and $a|c$, then $a|(bx + cy)$ for any integers $x, y$.

*Proof:* In each case the proof is simply a matter of going back to the definition of $a|b$. For instance, to prove (iv), the assumptions mean that there are integers $d$ and $e$ such that $b = da$ and $c = eb$, and it follows at once that $c = (de)a$, so $a|c$. To take another case, consider (vi). Since $a|b$, there is an integer $d$ such that $b = da$. Thus $|b| = |d| \cdot |a|$. Since $b \neq 0$, we must have $d \neq 0$ here, so $|d| \geq 1$. Thus $|a| \leq |b|$, as required. The remaining cases are left as an exercise. $\square$

## Exercises 4.1.3

   1. Prove all the parts of Theorem 4.1.3.

   2. Prove that every odd number is of one of the forms $4n + 1$ or $4n + 3$.

   3. Prove that for any integer $n$, at least one of the integers $n$, $n + 2$, $n + 4$ is divisible by 3.

   4. Prove that if $a$ is an odd integer, then $24|a(a^2 - 1)$. [Hint: Look at the example that followed Theorem 4.1.2.]

   5. Prove the following version of the Division Theorem. Given integers $a, b$ with $b \neq 0$, there are unique integers $q$ and $r$ such that

$$a = qb + r \quad \text{and} \quad -\tfrac{1}{2}|b| < r \leq \tfrac{1}{2}|b|$$

   [Hint: Write $a = q'b + r'$ where $0 \leq r' < |b|$. If $0 \leq r' \leq \tfrac{1}{2}|b|$, let $r = r', q = q'$. If $\tfrac{1}{2}|b| < r' < |b|$, let $r = r' - |b|$, and set $q = q' + 1$ if $b > 0$ and $q = q' - 1$ if $b < 0$.]

We have already met Euclid's proof that there are infinitely many primes. The formal definition of a *prime* number is an integer $p > 1$ which is only divisible by 1 and $p$.

A natural number $n > 1$ that is not prime is said to be *composite*.

## Exercises 4.1.4

1. Does the following statement accurately define prime numbers? Explain your answer. If the statement does not define the primes, modify it so it does.

$$p \text{ is prime iff } (\forall n \in \mathcal{N})[(n|p) \Rightarrow (n = 1 \lor n = p)]$$

2. A classic unsolved problem in number theory asks if there are infinitely many pairs of 'twin primes', pairs of primes separated by 2, such as 3 and 5, 11 and 13, or 71 and 73. Prove that the only prime triple (i.e. three primes, each 2 from the next) is 3, 5, 7.

3. It is a standard result about primes (known as Euclid's Lemma) that if $p$ is prime, then whenever $p$ divides a product $ab$, $p$ divides at least one of $a, b$. Prove the converse, that any natural number having this property (for any pair $a, b$) must be prime.

Most of the (considerable) interest in the prime numbers stems from their fundamental nature within the natural numbers, as expressed by the *Fundamental Theorem of Arithmetic*: Every natural numbers greater than 1 is either a prime or can be expressed as a product of prime numbers in a way that is unique except for the order in which they are written.

For example, 2, 3, 5, 7, 11, 13 are prime, and

$$
\begin{aligned}
4 &= 2 \times 2 = 2^2 \\
6 &= 2 \times 3 \\
8 &= 2 \times 2 \times 2 = 2^3 \\
9 &= 3 \times 3 = 3^2 \\
10 &= 2 \times 5 \\
12 &= 2 \times 2 \times 3 = 2^2 \cdot 3 \\
&\cdots \\
3366 &= 2 \cdot 3^2 \cdot 11 \cdot 17 \\
&\cdots
\end{aligned}
$$

The expression of a composite number as a product of primes is called its *prime decomposition*. Knowing the prime decomposition of a number tells you a lot about its mathematical properties. In this respect, the prime numbers are like the chemist's elements or the physicist's atoms.

Assuming Euclid's Lemma (the result mentioned in Exercise 4.1.4(3)), that if a prime $p$ divides a product $ab$, then $p$ divides at least one of $a, b$), we can prove the Fundamental Theorem of Arithmetic. (Euclid's Lemma is not particularly difficult, but it takes me beyond my goal of developing mathematical thinking.)

**Theorem 4.1.4 (Fundamental Theorem of Arithmetic)** Any natural number greater than 1 is either a prime or can be expressed as a product of prime numbers in a way that is unique except for the order in which they are written.

*Proof:* We prove the existence of a prime decomposition first. (This part does not require Euclid's Lemma.) I gave one proof in Chapter 3 as an illustration of induction. Here, I'll give a shorter proof by contradiction.

Assume there were a composite number that could not be written as a product of primes. Then there must be a smallest such number. Call it $n$. Since $n$ is not prime, there are numbers $a$, $b$, with $1 < a, b < n$, such that $n = a \cdot b$.

If $a$ and $b$ are primes, then $n = a \cdot b$ is a prime decomposition of $n$ and we have a contradiction.

If either of $a$, $b$ is composite, then because it is less than $n$, it must be a product of primes, so by replacing one or both of $a$, $b$ by their prime decompositions in $n = a \cdot b$, we obtain a prime decomposition of $n$, and again we have a contradiction.

We now turn to proving uniqueness. Again, we use the method of proof by contradiction. Assume there is a composite number that can be expressed in two genuinely different ways as products of primes. Let $n$ be the smallest such number, and let

$$(*) \qquad n = p_1 \cdot p_2 \cdot \ldots \cdot p_r = q_1 \cdot q_2 \cdot \ldots \cdot q_s$$

be two different prime decompositions of $n$.

Since $p_1$ divides $(q_1)(q_2 \cdot \ldots \cdot q_s)$ by Euclid's Lemma, either $p_1 | q_1$ or $p_1 | (q_2 \cdot \ldots \cdot q_s)$. Hence, either $p_1 = q_1$ or $p_1 = q_i$ for some $i$ between 2 and $s$, which can be expressed more concisely by saying that $p_1 = q_i$ for some $i$ between 1 and $s$. But then by removing $p_1$ and $q_i$ from the decompositions in $(*)$, we obtain a number smaller than $n$ having two different prime decompositions, which contradicts the choice of $n$ as the smallest such.

That completes the proof.     □

## Exercises 4.1.5

1. Try to prove Euclid's Lemma. If you do not succeed, move on to the following exercise.

2. You can find proofs of Euclid's Lemma in most textbooks on elementary number theory, and on the Web. Find a proof and make sure you understand it. If you find a proof on the Web, you will need to check that it is correct. There are false mathematical proofs all over the Internet. Though false proofs on Wikipedia usually get corrected fairly quickly, they also occasionally become corrupted when a well-intentioned contributor makes an attempted simplification that renders the proof incorrect. Learning how to make good use of Web resources is an important part of being a good mathematical thinker.

3. A fascinating and, it turns out, useful (both within mathematics and for real-world applications) result about prime numbers is *Fermat's Little Theorem*: If $p$ is prime and $a$ is a natural number that is not a multiple of $p$, then $p | (a^{p-1} - 1)$. Find (in a textbook or on the Web) and *understand* a proof of this result. (Again, be wary of mathematics you find on websites of unknown or non-accredited authorship.)

## 4.2   The real numbers

*If you are not familiar with elementary set theory, you should read the appendix before you proceed further in this chapter.*

Numbers arose from the formalization of two different human-cognitive conceptions: counting and measurement. Based on fossil records, anthropologists believe that both concepts existed and were used many thousands of years before numbers were introduced. As early as 35,000 years ago, humans put notches into bones (and probably wooden sticks as well, but none have survived and been found) to record things—possibly the cycles of the moon or the seasons—and it seems probable they used sticks or lengths of vine to measure lengths. Numbers themselves, however—abstractions that stand for the number of notches on a bone or the length of a measuring device—appear to have first appeared much later, around 10,000 years ago in the case of counting collections.

These activities resulted in two different kinds of number: the discrete counting numbers and the continuous real numbers. The connection between these two kinds of numbers was not finally put onto a firm footing until the nineteenth century, with the construction of the modern real number system. The reason it took so long is that the issues that had to be overcome were pretty subtle. Though the construction of the real numbers is beyond the scope of this book, I can explain what the problems were.

The connection between the two conceptions of numbers was made by showing how, starting with the integers $\mathcal{Z}$ (Z for *Zahlen*), it is possible to define first the rationals $\mathcal{Q}$ (Q for quotient) and then use the rationals to define the real numbers $\mathcal{R}$.

Starting with the integers, it is fairly straightforward to define the rational numbers. A rational number is, after all, simply a ratio of two integers. (It's actually not entirely trivial to construct the rational number system from the integers. Try your hand at the following exercise.)

*Exercise 4.2.1*

1. Take the integers, $\mathcal{Z}$, as a given system of numbers. You want to define a larger system, $\mathcal{Q}$, that extends $\mathcal{Z}$ by having a quotient $a/b$ for every pair $a$, $b$ of integers, $b \neq 0$. How would you go about defining such a system? In particular, how would you respond to the question, "What *is* the quotient $a/b$?" (You cannot answer in terms of actual quotients, since until $\mathcal{Q}$ has been defined, you don't have quotients.)

2. Find an account of the construction of the rationals from the integers and understand it, once again being cautious about mathematics found on the Internet.

With the rationals, you have a system of numbers adequate for any real-world measurement. This is captured by the following property of rational numbers.

**Theorem 4.2.1** If $r, s$ are rationals, $r < s$, then there is a rational $t$ such that $r < t < s$.

*Proof:* Let

$$t = \frac{1}{2}(r + s)$$

Clearly, $r < t < s$. But is $t \in \mathcal{Q}$? Well, letting $r = m/n$, $s = p/q$, where $m, n, p, q \in \mathcal{Z}$, we have

$$t = \frac{1}{2}\left(\frac{m}{n} + \frac{p}{q}\right) = \frac{mq + np}{2nq}$$

so as $mq + np, 2nq \in \mathcal{Z}$, we conclude that $t \in \mathcal{Q}$.  $\square$

The above property of there being a third rational between any two unequal rationals is known as *density*.

Because of density, for practical measurement in the world, the rational numbers are all you need. Between any two rational numbers there is a third. Hence, between any two rational numbers there are infinitely many other rational numbers. So you can measure anything in the world to whatever precision you need using the rational numbers.

But you need real numbers to do mathematics. The ancient Greeks learned that the rational numbers are not sufficient to provide (theoretical) *mathematical measurement* when they discovered that the length of the hypotenuse of a right-angled triangle with height and width equal to 1 unit is not a rational number. (The famous result that $\sqrt{2}$ is irrational, which we proved earlier.) This is not a problem for the construction engineer or the carpenter who has to work with right-angled triangles, but it is a major obstacle in mathematics itself.

The trouble is, although the rationals are dense (as defined above), there are nevertheless "holes" in the rational line. For example, if we let

$$\begin{aligned} A &= \{x \in \mathcal{Q} \mid x \le 0 \vee x^2 < 2\} \\ B &= \{x \in \mathcal{Q} \mid x > 0 \wedge x^2 \ge 2\} \end{aligned}$$

then every element of $A$ is less than every element of $B$, and

$$A \cup B = \mathcal{Q}$$

But $A$ has no greatest member and $B$ has no smallest member (as you can easily check), so there is a sort of hole between $A$ and $B$. This is the hole where $\sqrt{2}$ ought to be, of course. The fact that $\mathcal{Q}$ contains holes of this nature makes it unsuitable for some mathematical purposes, even though it suffices for all our measurements. Indeed, a number system in which the equation

$$x^2 - 2 = 0$$

has no solution is not going to support much advanced mathematics.

If the idea of there being holes in a densely packed line like the rational number line seems strange, things got even stranger when mathematicians finally did figure out how to fill those holes. The numbers that fill in the holes are known as *irrational numbers*. Taken together, the rational numbers and the irrational numbers constitute what are called the *real numbers*. It turns out that when you fill in the holes in the rational line, you get a lot more numbers than you bargained for. Not only are there infinitely many irrational numbers between any two rational numbers, but in a very precise sense there are "infinitely more" irrational numbers between them then there are rational numbers between them. The irrational numbers so dominate the real line that if you were to select a real number at random, the mathematical probability that it would be irrational is 1.

There are several different ways to construct the real numbers from the rational numbers in a rigorous fashion, all of which would take us beyond our present scope. But at an intuitive level, the idea is to allow decimal expansions to be infinite. In the case of infinite recurring decimals, the expression denotes a rational number, such as 0.333 . . ., which is 1/3, or 0.142857 142857 142857 . . ., which is 1/7. But if there is no recurring pattern, the result is an irrational number, for example $\sqrt{2}$ starts off 1.41421356237309504880168872420969807 . . . and continues forever without settling into a recurring pattern.

## 4.3 Completeness

One of the most valuable results to come out of the construction of the real number system was the formulation of a simple property of the reals that captures those infinitesimal holes in the rational line and specifies exactly how they are filled. It is called the *Completeness Property*. Before I can explain it, we need some familiarity with the real line as an ordered set.

Certain types of subset of the reals $\mathcal{R}$ occur so frequently that it is convenient to introduce a special notation for them.

By an *interval* we mean an uninterrupted stretch of the real line. There are a number of different kinds of interval, for which there is a fairly widespread standard notation.

Let $a, b \in \mathcal{R}, a < b$. The *open interval* $(a, b)$ is the set

$$(a, b) = \{x \in \mathcal{R} \mid a < x < b\}$$

The *closed interval* $[a, b]$ is the set

$$[a, b] = \{x \in \mathcal{R} \mid a \leq x \leq b\}$$

The point to notice here is that neither $a$ nor $b$ is an element of $(a, b)$, but both $a$ and $b$ are elements of $[a, b]$. (This seemingly trivial distinction turns out to be highly significant in elementary real analysis.) Thus, $(a, b)$ is the stretch of the real line beginning 'just past' $a$ and ending 'just before' $b$, while $[a, b]$ is the stretch beginning with $a$ and ending with $b$.

The above notation extends in an obvious manner. We call

$$[a, b) = \{x \in \mathcal{R} \mid a \leq x < b\}$$

a *left-closed, right-open* interval, and

$$(a, b] = \{x \in \mathcal{R} \mid a < x \leq b\}$$

a *left-open, right-closed* interval.

Both $[a, b)$ and $(a, b]$ are sometimes referred to as *half-open* (or *half-closed*) intervals.

Finally, we set

$$\begin{array}{rcl}
(-\infty, a) & = & \{x \in \mathcal{R} \mid x < a\} \\
(-\infty, a] & = & \{x \in \mathcal{R} \mid x \leq a\} \\
(a, \infty) & = & \{x \in \mathcal{R} \mid x > a\} \\
[a, \infty) & = & \{x \in \mathcal{R} \mid x \geq a\}
\end{array}$$

Notice that the symbol $\infty$ is never coupled with a square bracket. This would be misleading, since $\infty$ is *not* a number, just a useful symbol. In the above definitions it simply helps us to extend a convenient notation to cover another case.

## *Exercises 4.3.1*

1. Prove that the intersection of two intervals is again an interval. Is the same true for unions?

2. Taking $\mathcal{R}$ as the universal set, express the following as simply as possible in terms of intervals and unions of intervals. (Note that $A'$ denotes the complement of the set $A$ relative to the given universal set, which in this case is $\mathcal{R}$. See the appendix.)

   (a) $[1, 3]'$                      (b) $(1, 7]'$
   (c) $(5, 8]'$                     (d) $(3, 7) \cup [6, 8]$
   (e) $(-\infty, 3)' \cup (6, \infty)$       (f) $\{\pi\}'$
   (g) $(1, 4] \cap [4, 10]$            (h) $(1, 2) \cap [2, 3)$
   (i) $A'$, where $A = (6, 8) \cap (7, 9]$    (j) $A'$, where $A = (-\infty, 5] \cup (7, \infty)$

Now we are in a position to take a look at the way the modern real number system handles the notion of "filling in the holes" of the rational line.

Given a set $A$ of reals, a number $b$ such that $(\forall a \in A)[a \leq b]$ is said to be an *upper bound* of $A$.

We say $b$ is a *least upper bound* of $A$ if, in addition, for any upper bound $c$ of $A$, we have $b \leq c$.

The same definitions can be made for sets of integers or sets of rationals, of course.

The least upper bound of a set $A$ is often written $\mathrm{lub}(A)$.

*The Completeness Property* of the real number system says that any nonempty set of reals that has an upper bound has a least upper bound (in $\mathcal{R}$).

*Exercises 4.3.2*

1. Prove that if a set $A$ of integers/rationals/reals has an upper bound, then it has infinitely many different upper bounds.

2. Prove that if a set $A$ of integers/rationals/reals has a least upper bound, then it is unique.

3. Let $A$ be a set of integers, rationals, or reals. Prove that $b$ is the least upper bound of $A$ iff:

   (a) $(\forall a \in A)(a \leq b)$; and

   (b) whenever $c < b$ there is an $a \in A$ such that $a > c$.

4. The following variant of the above characterization is often found. Show that $b$ is the lub of $A$ iff:

   (a) $(\forall a \in A)(a \leq b)$ ; and

   (b) $(\forall \epsilon > 0)(\exists a \in A)(a > b - \epsilon)$.

5. Give an example of a set of integers that has no upper bound.

6. Show that any finite set of integers/rationals/reals has a least upper bound.

7. Intervals: What is lub $(a, b)$? What is lub $[a, b]$? What is max $(a, b)$? What is max $[a, b]$?

8. Let $A = \{|x - y| \mid x, y \in (a, b)\}$. Prove that $A$ has an upper bound. What is lub $A$ ?

9. Define the notion of a *lower bound* of a set of integers/rationals/reals.

10. Define the notion of a *greatest lower bound* (glb) of a set of integers/rationals/reals by analogy with our original definition of lub.

11. State and prove the analog of Question 3 for greatest lower bounds.

12. State and prove the analog of Question 4 for greatest lower bounds.

13. Show that the Completeness Property for the real number system could equally well have been defined by the statement, "Any nonempty set of reals that has a lower bound has a greatest lower bound."

14. The integers satisfy the Completeness Property, but for a trivial reason. What is that reason?

**Theorem 4.3.1** The rational line, $\mathcal{Q}$, does not have the completeness property.

*Proof:* Let

$$A = \{r \in \mathcal{Q} \mid r \geq 0 \wedge r^2 < 2\}$$

$A$ is bounded above in $\mathcal{Q}$ by 2. But it has no least upper bound in $\mathcal{Q}$. Intuitively, this is because the only possible least upper bound would be $\sqrt{2}$, which we know is not in $\mathcal{Q}$, but we shall prove this rigorously.

Let $x \in \mathcal{Q}$ be any upper bound of $A$. We show that there is a smaller one (in $\mathcal{Q}$).

Let $x = p/q$, where $p, q \in \mathcal{N}$.

Suppose first that $x^2 < 2$. Thus $2q^2 > p^2$. Now, as $n$ gets large, the expression $n^2/(2n+1)$ increases without bound, so we can pick $n \in \mathcal{N}$ so large that

$$\frac{n^2}{2n+1} > \frac{p^2}{2q^2 - p^2}$$

Rearranging, this gives

$$2n^2 q^2 > (n+1)^2 p^2$$

Hence

$$\left(\frac{n+1}{n}\right)^2 \frac{p^2}{q^2} < 2$$

Let

$$y = \left(\frac{n+1}{n}\right)\frac{p}{q}$$

Thus $y^2 < 2$. Now, since $(n+1)/n > 1$, we have $x < y$. But $y$ is rational and we have just seen that $y^2 < 2$, so $y \in A$. This contradicts the fact that $x$ is an upper bound for $A$.

It follows that we must have $x^2 \geq 2$. Since there is no rational whose square is equal to 2, this means that $x^2 > 2$. Thus $p^2 > 2q^2$, and we can pick $n \in \mathcal{N}$ so large now that

$$\frac{n^2}{2n+1} > \frac{2q^2}{p^2 - 2q^2}$$

which becomes, upon rearranging,

$$p^2 n^2 > 2q^2(n+1)^2$$

i.e.

$$\frac{p^2}{q^2}\left(\frac{n}{n+1}\right)^2 > 2$$

Let

$$y = \left(\frac{n}{n+1}\right)\frac{p}{q}$$

Then $y^2 > 2$. Since $n/(n+1) < 1$, $y < x$. But for any $a \in A$, $a^2 < 2 < y^2$, so $a < y$. Thus $y$ is an upper bound of $A$ less than $x$, as we set out to prove.   $\square$

*Exercises 4.3.3*

1. Let $A = \{r \in \mathcal{Q} \mid r > 0 \wedge r^2 > 3\}$. Show that $A$ has a lower bound in $\mathcal{Q}$ but no greatest lower bound in $\mathcal{Q}$. Give all details of the proof along the lines of Theorem 4.3.1.

2. In addition to the completeness property, the *Archimedean property* is an important fundamental property of $\mathcal{R}$. It says is that if $x, y \in \mathcal{R}$ and $x, y > 0$, there is an $n \in \mathcal{N}$ such that $nx > y$.

   Use the Archimedean property to show that if $r, s \in \mathcal{R}$ and $r < s$, there is a $q \in \mathcal{Q}$ such that $r < q < s$. (Hint: pick $n \in \mathcal{N}$, $n > 1/(s - r)$, and find an $m \in \mathcal{N}$ such that $r < (m/n) < s$.)

# 4.4   Sequences

Suppose we associate with each natural number $n$ a real number $a_n$. The set of all these numbers $a_n$, arranged according to the index $n$, is called a *sequence*. We denote this sequence by

$$\{a_n\}_{n=1}^{\infty}$$

Thus, the symbol $\{a_n\}_{n=1}^{\infty}$ represents the sequence

$$a_1, a_2, a_3, \ldots, a_n, \ldots$$

For example, the members of $\mathcal{N}$ themselves constitute a sequence when assigned their usual order

$$1, 2, 3, \ldots, n, \ldots$$

This sequence would be denoted by $\{n\}_{n=1}^{\infty}$ (because $a_n = n$ for each $n$).
   Or we could order the elements of $\mathcal{N}$ in a different manner to obtain the sequence

$$2, 1, 4, 3, 6, 5, 8, 7, \ldots$$

This is quite a different sequence from the sequence $\{n\}_{n=1}^{\infty}$, since the ordering in which the members of the sequence appear is important. Or, if we allow repetitions we get a completely new sequence

$$1, 1, 2, 2, 3, 3, 4, 4, 4, 5, 6, 7, 8, 8, \ldots$$

There does not need to be a nice rule involved. It may be impossible to find a formula to describe $a_n$ in terms of $n$, though the specific examples you find in textbooks do, of course, have rules.
   Again, we can have a constant sequence

$$\{\pi\}_{n=1}^{\infty} = \pi, \pi, \pi, \pi, \pi, \ldots, \pi, \ldots$$

or an alternating (in sign) sequence

$$\{(-1)^{n+1}\}_{n=1}^{\infty} = +1, -1, +1, -1, +1, -1, \ldots$$

In short, there is no restriction on what the members of a sequence $\{a_n\}_{n=1}^{\infty}$ may be, except that they be real numbers.

Some sequences have a rather special property. As you go along the sequence, the numbers in the sequence get arbitrarily closer and closer to some fixed number. For instance, the members of the sequence

$$\left\{\frac{1}{n}\right\}_{n=1}^{\infty} = 1, \tfrac{1}{2}, \tfrac{1}{3}, \tfrac{1}{4}, \ldots, \tfrac{1}{n}, \ldots$$

get arbitrarily closer and closer to 0 as $n$ gets larger, and the members of the sequence

$$\left\{1 + \frac{1}{2^n}\right\}_{n=1}^{\infty} = 1\tfrac{1}{2}, 1\tfrac{1}{4}, 1\tfrac{1}{8}, 1\tfrac{1}{16}, \ldots$$

get arbitrarily closer and closer to 1. Again, the members of the sequence

$$3,\ 3.1,\ 3.14,\ 3.141,\ 3.1415,\ 3.14159,\ 3.141592,\ 3.1415926,\ \ldots$$

get arbitrarily closer and closer to $\pi$, although this example is not as good as some of the others, since we have not given a general rule for the $n$th term in the sequence.

If the members of the sequence $\{a_n\}_{n=1}^{\infty}$ get arbitrarily closer and closer to some fixed number $a$ in this manner, we say that the sequence $\{a_n\}_{n=1}^{\infty}$ *tends to the limit* $a$, and write

$$a_n \to a \quad \text{as} \quad n \to \infty$$

Another common notation is

$$\lim_{n \to \infty} a_n = a$$

So far, this is all at an intuitive level. Let us see if we can obtain a precise definition of what it means to write "$a_n \to a$ as $n \to \infty$".

Well, to say that $a_n$ gets arbitrarily closer and closer to $a$ is to say that the difference $|a_n - a|$ gets arbitrarily closer and closer to 0. This is the same as saying that whenever $\epsilon$ is a positive real number, the difference $|a_n - a|$ is eventually less than $\epsilon$. This leads to the following formal definition:

$$a_n \to a \quad \text{as} \quad n \to \infty \quad \text{iff}$$

$$(\forall \epsilon > 0)(\exists n \in \mathcal{N})(\forall m \geq n)(|a_m - a| < \epsilon)$$

This looks quite complicated. Let us try to analyze it. Consider the part

$$(\exists n \in \mathcal{N})(\forall m \geq n)(|a_m - a| < \epsilon)$$

This says that there is an $n$ such that for all $m$ greater than or equal to $n$, the distance from $a_m$ to $a$ is less than $\epsilon$. In other words, there is an $n$ such that all terms in the sequence $\{a_n\}_{n=1}^{\infty}$ beyond $a_n$ lie within the distance $\epsilon$ of $a$. We can express this concisely by saying that the terms in the sequence $\{a_n\}_{n=1}^{\infty}$ are eventually all within the distance $\epsilon$ from $a$.

Thus, the statement

$$(\forall \epsilon > 0)(\exists n \in \mathcal{N})(\forall m \geq n)(|a_m - a| < \epsilon)$$

says that for every $\epsilon > 0$, the members of the sequence $\{a_n\}_{n=1}^{\infty}$ are eventually all within the distance $\epsilon$ from $a$. This is the formal definition of the intuitive notion of "$a_n$ gets arbitrarily closer and closer to $a$".

Let us consider a numerical example. Consider the sequence $\{1/n\}_{n=1}^{\infty}$. On an intuitive level, we know that $1/n \to 0$ as $n \to \infty$. We shall see how the formal definition works for this sequence. We must prove that

$$(\forall \epsilon > 0)(\exists n \in \mathcal{N})(\forall m \geq n)(|\tfrac{1}{m} - 0| < \epsilon)$$

This simplifies at once to

$$(\forall \epsilon > 0)(\exists n \in \mathcal{N})(\forall m \geq n)(\tfrac{1}{m} < \epsilon)$$

To prove that this is a true assertion, let $\epsilon > 0$ be arbitrary. We must find an $n$ such that

$$m \geq n \Rightarrow \tfrac{1}{m} < \epsilon$$

Pick $n$ large enough so that $n > 1/\epsilon$. (This uses the Archimedean property of $\mathcal{R}$ discussed in Exercises 4.3.) If now $m \geq n$ then

$$\frac{1}{m} \leq \frac{1}{n} < \epsilon$$

In other words,

$$(\forall m \geq n)(\tfrac{1}{m} < \epsilon)$$

as required.

One point to notice here is that our choice of $n$ depended upon the value of $\epsilon$. The smaller $\epsilon$ is, the greater our $n$ needs to be.

Another example is the sequence $\{n/(n+1)\}_{n=1}^{\infty}$, i.e.

$$\frac{1}{2}, \frac{2}{3}, \frac{3}{4}, \frac{4}{5}, \ldots$$

We prove that $n/(n+1) \to 1$ as $n \to \infty$. Let $\epsilon > 0$ be given. We must find an $n \in \mathcal{N}$ such that for all $m \geq n$

$$\left| \frac{m}{m+1} - 1 \right| < \epsilon$$

Pick $n$ so large that $n > 1/\epsilon$. Then, for $m \geq n$,

$$\left| \frac{m}{m+1} - 1 \right| = \left| \frac{-1}{m+1} \right| = \frac{1}{m+1} < \frac{1}{m} \leq \frac{1}{n} < \epsilon$$

as required.

## Exercises 4.4.1

(1) Formulate both in symbols and in words what it means to say that $a_n \not\to a$ as $n \to \infty$.

(2) Prove that $(n/(n+1))^2 \to 1$ as $n \to \infty$.

(3) Prove that $1/n^2 \to 0$ as $n \to \infty$.

(4) Prove that $1/2^n \to 0$ as $n \to \infty$.

(5) We say a sequence $\{a_n\}_{n=1}^{\infty}$ *tends to infinity* if, as $n$ increases, $a_n$ increases without bound. For instance, the sequence $\{n\}_{n=1}^{\infty}$ tends to infinity, as does the sequence $\{2^n\}_{n=1}^{\infty}$. Formulate a precise definition of this notion, and prove that both of these examples fulfill the definition.

(6) Let $\{a_n\}_{n=1}^{\infty}$ be an increasing sequence (i.e. $a_n < a_{n+1}$ for each $n$). Suppose that $a_n \to a$ as $n \to \infty$. Prove that $a = \text{lub}\{a_1, a_2, a_3, \ldots\}_{n=1}^{\infty}$.

(7) Prove that if $\{a_n\}_{n=1}^{\infty}$ is increasing and bounded above, then it tends to a limit.

# APPENDIX: Set theory

Most readers of this book will have learned enough basic set theory. This brief appendix summarizes what is required.

The concept of a *set* is extremely basic and pervades the whole of present-day mathematical thought. Any well-defined collection of objects is a set. For instance we have:

- the set of all students in your class

- the set of all prime numbers

- the set whose only member is you.

All it takes to determine a set is some way of specifying the collection. (Actually, that is not correct. In the mathematical discipline called abstract set theory, arbitrary collections are allowed, where there is no defining property.)

If $A$ is a set, then the objects in the collection $A$ are called either the *members* of $A$ or the *elements* of $A$. We write

$$x \in A$$

to denote that $x$ is an element of $A$.

Some sets occur frequently in mathematics, and it is convenient to adopt a standard notation for them:

$\mathcal{N}$ : the set of all natural numbers (i.e., the numbers 1, 2, 3, etc.)

$\mathcal{Z}$ : the set of all integers (0 and all positive and negative whole numbers)

$\mathcal{Q}$ : the set of all rational numbers (fractions)

$\mathcal{R}$ : the set of all real numbers

Thus, for example,

$$x \in \mathcal{R}$$

means that $x$ is a real number. And

$$(x \in \mathcal{Q}) \wedge (x > 0)$$

means that $x$ is a positive rational number.

There are several ways of specifying a set. If it has a small number of elements, we can list them. In this case, we denote the set by enclosing the list of the elements in curly brackets; thus, for example,

$$\{1, 2, 3, 4, 5\}$$

denotes the set consisting of the natural numbers 1, 2, 3, 4 and 5.

By use of 'dots' we can extend this notation to any finite set; e.g.,

$$\{1, 2, 3, \ldots, n\}$$

denotes the set of the first $n$ natural numbers. Again

$$\{2, 3, 5, 7, 11, 13, 17, \ldots, 53\}$$

could (given the right context) be used to denote the set of all primes up to 53.

Certain infinite sets can also be described by the use of dots (only now the dots have no end), e.g.,

$$\{2, 4, 6, 8, \ldots, 2n, \ldots\}$$

denotes the set of all even natural numbers. Again,

$$\{\ldots, -8, -6, -4, -2, 0, 2, 4, 6, 8, \ldots\}$$

denotes the set of all even integers.

In general, however, except for finite sets with only a small number of elements, sets are best described by giving the property which defines the set. If $A(x)$ is some property, the set of all those $x$ which satisfy $A(x)$ is denoted by

$$\{x \mid A(x)\}$$

Or, if we wish to restrict the $x$ to those which are members of a certain set $X$, we would write

$$\{x \in X \mid A(x)\}$$

This is read "the set of all $x$ in $X$ such that $A(X)$". For example:

$$
\begin{aligned}
\mathcal{N} &= \{x \in \mathcal{Z} \mid x > 0\} \\
\mathcal{Q} &= \{x \in \mathcal{R} \mid (\exists m, n \in \mathcal{Z})[(m > 0) \wedge (mx = n)]\} \\
\{\sqrt{2}, -\sqrt{2}\} &= \{x \in \mathcal{R} \mid x^2 = 2\} \\
\{1, 2, 3\} &= \{x \in \mathcal{N} \mid x < 4\}
\end{aligned}
$$

Two sets, $A, B$ are *equal*, written $A = B$, if they have exactly the same elements. As the above example shows, equality of sets does not mean they have identical definitions; there are often many different ways of describing the same set. The definition of equality reflects rather the fact that a set is just a collection of objects.

If we have to prove that the sets $A$ and $B$ are equal, we usually split the proof into two parts:

(a) Show that every member of $A$ is a member of $B$.

(b) Show that every member of $B$ is a member of $A$.

Taken together, (a) and (b) clearly imply $A = B$. (The proof of both (a) and (b) is usually of the 'take an arbitrary element' variety. To prove (a), for instance, we must prove $(\forall x \in A)(x \in B)$; so we take an arbitrary element $x$ of A and show that $x$ must be an element of $B$.)

The set notations introduced have obvious extensions. For instance, we can write

$$\mathcal{Q} = \{m/n \mid m, n \in \mathcal{Z}, n \neq 0\}$$

and so on.

It is convenient in mathematics to introduce a set which has no elements: the *empty set* (or *null set*). There will only be one such set, of course, since any two such will have exactly the same elements and thus be (by definition) equal. The empty set is denoted by the Scandinavian letter $\emptyset$ (Note that this is not the Greek letter $\phi$.) The empty set can be specified in many ways; e.g.,

$$\emptyset \;=\; \{x \in \mathcal{R} \mid x^2 < 0\}$$
$$\emptyset \;=\; \{x \in \mathcal{N} \mid 1 < x < 2\}$$
$$\emptyset \;=\; \{x \mid x \neq x\}$$

Notice that $\emptyset$ and $\{\emptyset\}$ are quite different sets. $\emptyset$ is the empty set: it has NO members. $\{\emptyset\}$ is a set which has ONE member. Hence

$$\emptyset \neq \{\emptyset\}$$

What is the case here is that

$$\emptyset \in \{\emptyset\}$$

(The fact that the single element of $\{\emptyset\}$ is the empty set is irrelevant in this connection: $\{\emptyset\}$ *does* have an element, $\emptyset$ does *not*.)

A set $A$ is called a *subset* of a set $B$ if every element of $A$ is a member of $B$. For example, $\{1, 2\}$ is a subset of $\{1, 2, 3\}$. We write

$$A \subseteq B$$

to mean that $A$ is a subset of $B$. If we wish to emphasize that $A$ and $B$ are unequal here, we write

$$A \subset B$$

and say that $A$ is a *proper subset* of $B$ (This usage compares with the ordering relations $\leq$ and $<$ on $\mathcal{R}$.)

Clearly, for any sets $A, B$, we have

$$A = B \;\; \text{iff} \;\; (A \subseteq B) \wedge (B \subseteq A)$$

## Exercises A1

1. What well-known set is this:

$$\{n \in \mathcal{N} \mid (n > 1) \wedge (\forall x, y \in \mathcal{N})[(xy = n) \Rightarrow (x = 1 \vee y = 1)]\}$$

2. Let

$$P = \{x \in \mathcal{R} \mid \sin(x) = 0\} \quad , \quad Q = \{n\pi \mid n \in \mathcal{Z}\}$$

What is the relationship between $P$ and $Q$ ?

3. Let
$$A = \{x \in \mathcal{R} \mid (x > 0) \wedge (x^2 = 3)\}$$

Give a simpler definition of the set $A$.

4. Prove that for any set $A$:

$$\emptyset \subseteq A \quad \text{and} \quad A \subseteq A$$

5. Prove that if $A \subseteq B$ and $B \subseteq C$, then $A \subseteq C$

6. List all subsets of the set $\{1, 2, 3, 4\}$.

7. List all subsets of the set $\{1, 2, 3, \{1, 2\}\}$.

8. Let $A = \{x \mid P(x)\}, B = \{x \mid Q(x)\}$, where $P, Q$ are formulas such that $\forall x[P(x) \Rightarrow Q(x)]$. Prove that $A \subseteq B$.

9. Prove (by induction) that a set with exactly $n$ elements has $2^n$ subsets.

10. Let
$$A = \{o, t, f, s, e, n\}$$

Give an alternative definition of the set $A$. (Hint: this is connected with $\mathcal{N}$ but is not entirely mathematical.)

There are various natural operations we can perform on sets. (They correspond *roughly* to addition, multiplication, and negation for integers.)

Given two sets $A, B$ we can form the set of all objects which are members of either one of $A$ and $B$. This set is called the *union* of $A$ and $B$ and is denoted by

$$A \cup B$$

Formally, this set has the definition

$$A \cup B = \{x \mid (x \in A) \vee (x \in B)\}$$

(Note how this is consistent with our decision to use the word 'or' to mean inclusive-or.)

The *intersection* of the sets $A, B$ is the set of all members which $A$ and $B$ have in common. It is denoted by

$$A \cap B$$

and has the formal definition

$$A \cap B = \{x \mid (x \in A) \wedge (x \in B)\}$$

Two sets $A, B$ are said to be *disjoint* if they have no elements in common: that is, if $A \cap B = \emptyset$.

The set-theoretic analog of negation requires the concept of a *universal set*. Often, when we are dealing with sets, they all consist of objects of the same kind. For example, in number theory we may focus on sets of natural numbers or sets of rationals; in real analysis we usually focus on sets of reals. A *universal set* for a particular discussion is simply the set of all objects of the kind being considered. It is frequently the domain over which the quantifiers range.

Once we have fixed a universal set we can introduce the notion of the *complement* of the set $A$. Relative to the universal set $U$, the complement of a set $A$ is the set of all elements of $U$ that are not in $A$. This set is denoted by $A'$, and has the formal definition

$$A' = \{x \in U \mid x \notin A\}$$

[Notice that we write $x \notin A$ instead of $\neg(x \in A)$, for brevity.]

For instance, if the universal set is the set $\mathcal{N}$ of natural numbers, and $E$ is the set of even (natural) numbers, then $E'$ is the set of odd (natural) numbers.

The following theorem sums up the basic facts about the three set operations just discussed.

**Theorem** Let $A, B, C$ be subsets of a universal set $U$.

(1) $A \cup (B \cup C) = (A \cup B) \cup C$

(2) $A \cap (B \cap C) = (A \cap B) \cap C$
    ((1) and (2) are the associative laws)

(3) $A \cup B = B \cup A$

(4) $A \cap B = B \cap A$
    ((3) and (4) are the commutative laws)

(5) $A \cup (B \cap C) = (A \cup B) \cap (A \cup C)$

(6) $A \cap (B \cup C) = (A \cap B) \cup (A \cap C)$
    ((5) and (6) are the distributive laws)

(7) $(A \cup B)' = A' \cap B'$

(8) $(A \cap B)' = A' \cup B'$
    ((7) and (8) are called the De Morgan laws)

(9) $A \cup A' = U$

(10) $A \cap A' = \emptyset$

    ((9) and (10) are the complementation laws)

(11) $(A')' = A$

    (self-inverse law)

*Proof:* Left as an exercise.   □

## Exercises A2

1. Prove all parts of the above theorem.

2. Find a resource that explains *Venn diagrams* and use them to illustrate and help you understand the above theorem.

# Index